SpringerBriefs in Operations Research

More information about this series at http://www.springer.com/series/11467

Maria Jesus Saenz · Eduardo Ubaghs
Alejandra Isabel Cuevas

Enabling Horizontal Collaboration Through Continuous Relational Learning

Springer

Maria Jesus Saenz
Zaragoza Logistics Center
Zaragoza
Spain

Alejandra Isabel Cuevas
Hewlett-Packard
Zaragoza
Spain

Eduardo Ubaghs
Thermal Services
Alstom (Switzerland) Ltd.
Baden
Switzerland

ISSN 2195-0482 ISSN 2195-0504 (electronic)
ISBN 978-3-319-08092-5 ISBN 978-3-319-08093-2 (eBook)
DOI 10.1007/978-3-319-08093-2

Library of Congress Control Number: 2014946376

Springer Cham Heidelberg New York Dordrecht London

Printed on acid-free paper

Springer is part of Springer Science+Business Media (www.springer.com)

Contents

Abstract

Now that supply chain operations are facing new challenges due to the necessity of reducing costs and CO_2 emissions, companies are increasingly leveraging cooperation with companies from other supply chains as a source of competitive advantage. The aim of this book is to build upon a framework of inter-organizational learning for horizontal collaboration, and the outputs from this study are designed to help in understanding the relationship dynamics and the managerial insights of horizontal collaboration for sustainable and profitable cooperation in freight distribution. The proposed framework is based on exploratory research, including semi-structured interviews with senior managers from four different companies from the home appliances, automotive, retail, and food and beverage industries. Additionally, collaborative evidence from different successful pilot cases relevant to the distribution operations of the four companies are described in order to illustrate the framework presented. The study demonstrates that by applying horizontal collaboration practices companies can learn and create absorptive capacity. Our case studies provide in-depth explanation of how the relational elements of cooperation in freight distribution can lead absorptive capacity toward overcoming the barriers to collaboration and toward efficiency, sustainability, and innovation. The successful pilot cases demonstrate the following benefits. (a) In terms of efficiency, intelligent freight combinations have achieved savings between 9 and 49 %, depending on the different applications. (b) In terms of effectiveness, horizontal collaboration partnerships have allowed incremental delivery frequencies and service level. (c) Regarding sustainability, consortiums have achieved between 20 and 30 % savings in CO_2 emissions. (d) Supply chain managers are able to heed all of the elements that will help their companies improve their performance through horizontal collaboration. This book fills a gap in the literature by providing a generic model for horizontal collaboration dynamics. It can be used by any participant involved in or willing to start horizontal collaboration in freight distribution.

Keywords Horizontal collaboration · Inter-organizational learning · Collaborative relationships · Supply chain management · Successful case studies · Freight distribution

Chapter 1
Introduction

The situation of the European freight-transportation industry is critical. A report of the World Economic Forum (WEF) presented disturbing numbers: 24 % of vehicles run empty, and the rest are only 57 % full, which results in an overall capacity utilization of only 43 %. The abatement potential in terms of both cost and carbon is vast. For example, increasing the load factor up to 70 % would generate an estimated savings of 22 Billion euros and a reduction of 61 million tons of CO_2 emissions (WEF 2009). In a time when both the world economy and environment are facing serious challenges, European companies should react by replacing their traditional procurement and delivery methods with more innovative and efficient distribution practices. As stated in a recent study conducted by McKinsey & Co., to date companies have mainly selected an internal approach to supply chain decarbonisation. A survey included in the same report shows that despite increasing supply chain collaborations in Europe this effort still has enormous upside potential. About 60 % of the respondents stated that their collaborative initiatives had led to substantial benefits, and optimising supply chain flows and processes through collaboration decreased these companies' costs by up to 51 % (Hua et al. 2012).

There is significant room for the companies to exploit these efficiency gaps, but the question is how to do it. Many key policy makers and industrial players in the logistics community have recognized that the key strategy for overcoming this situation is to stimulate the collaboration between industry players in the distribution of goods, which has recently been called "horizontal collaboration."

In order to address these challenges, the European Commission has launched a research initiative called Collaboration Concepts for Co-modality (CO^3), funded by the Seventh Framework Programme (grant agreement 284926). The mission of CO^3 is to encourage a structural breakthrough in the competitiveness and sustainability of European logistics by stimulating horizontal collaboration among European shipping companies (www.co3-project.eu). The CO^3 consortium is made up of logistics and manufacturing—industry specialists and transport service providers. Several research institutions, such as the Zaragoza Logistics Center, work together

M.J. Saenz et al., *Enabling Horizontal Collaboration Through Continuous Relational Learning*, SpringerBriefs in Operations Research,
DOI 10.1007/978-3-319-08093-2_1

to create business models for horizontal collaboration. Additionally, a High-Level Industry Board representing companies such as P&G, Nike, Kimberly Clark, Unilever, HJ Heinz, Nestlé, DHL, and Kuehne Nagel, among others, provides business expertise and feedback for the applicability of the business models developed through the CO^3 project.

Within this context, the contribution of the research group at the Zaragoza Logistics Center (www.zlc.edu.es), which is a research institute established by the Government of Aragon in Spain in partnership with the Massachusetts Institute of Technology (U.S.) and the University of Zaragoza is aimed at providing the elements for successful business models and the corresponding inter-organizational learning dynamics for horizontal collaboration that will enable long-term, sustainable cooperation.

The present book starts with a review of the literature on inter-organizational learning, followed by an elaboration of the connection between vertical and horizontal collaboration. Whereas vertical cooperation has been widely exploited and analysed, the literature on horizontal collaboration in logistics is still in its infancy (Cruijssen et al. 2007). However, this kind of collaborative practice is gaining momentum within the transportation sector (Cruijssen et al. 2006). Cooperating with partners from other supply chains with common networks and complementary needs and operations can increase companies' capacities to respond to certain operational concerns, such as fluctuating demand (Mason et al. 2007), and boost distribution efficiency by decreasing costs, improving service, and protecting market positions. The 40 % of the respondents from the McKinsey & Co. survey that did not perceive any positive impact from collaboration argued that the reasons for failure were the lack of support from top management, insufficient resources and effort, and resistance to sharing information (Hua et al. 2012). Lyons et al. (2012) undertook a survey about horizontal collaboration drivers and barriers, in which a lack of trust and the fear of competitors gaining sensitive company information were major obstacles to implementing horizontal collaboration practices. Thus, building up a trusting relationship and sharing sensitive information are essential aspects to be settled when facing collaboration because the company's resources will span its own boundaries to merge with the ones from its partner (Dyer and Singh 1998).

This constitutes a context in which the firms will have to learn from each other in order to gain the potential synergies behind these horizontal collaboration opportunities. This requires establishing the right conditions in the collaborative dyad or network for continuously exploring new ideas and principles, engaging in information sharing, and structuring logistical flows and operations in order to allow the realization of relational rents. These learning mechanisms are facilitated by the creation of AC (Absorptive Capacity) in order to generate common and sustainable benefits. Absorptive capacity in a relational context is the stock of knowledge that results from flows of knowledge across organizations, and it emerges through interactions over long periods of time in the relationship (Sáenz et al. 2014). Lane et al. (2006) propose that the latter has three stages, with each stage corresponding to a different learning process: exploration, assimilation, and exploitation.

Based on these dynamics, the aim of this book is to set up a model that enables horizontal collaboration by overcoming the barriers to collaboration and building up the foundations for sustainable and profitable cooperation in freight distribution. The different sections analyse the elements that are needed for continuous relational learning that facilitate mature levels in a horizontal collaborative relationship. This research also analyses the various expected outcomes, after activating all these elements, in terms of the relational rents, such as efficiency, operational flexibility, service level, collaborative innovation, and sustainability.

The model presented is empirically validated by semi-structured interviews with senior managers from four different companies from the home appliances, automotive, retail, and food and beverages industries. It is finally complemented with the experiences and collaborative evidence from different pilot cases about the distribution operations of various horizontal-collaboration consortiums. Finally, we provide the derived managerial recommendations and the main conclusions from the study, and we propose future research directions.

References

CO^3 Project. Collaboration Concepts for Co-modality (Seventh Framework Programme, Ref. 284926). www.co3-project.eu

Cruijssen F, Dullaert W, Joro T (2006) Logistics efficiency through horizontal cooperation: the case of Flemish road transportation companies. CentER discussion paper

Cruijssen F, Dullaert W, Fleuren H (2007) Horizontal cooperation in transport and logistics: a literature review. Transp J 46(3):22–39

Dyer JH, Singh H (1998) The relational view: cooperative strategy and sources of interorganizational competitive advantage. Acad Manag Rev 23(4):660–679

Hua X, Brinkhoff A, Großpietsch J, Losdat C, Nyssens JA, Weng J (2012) Collaboration today and tomorrow-taking the right steps on the path to impact. ECR Europe/McKinsey & Company survey

Lane PJ, Koka BR, Pathak S (2006) The reification of absorptive capacity: a critical review and rejuvenation of the construct. Acad Manag Rev 31(4):833–863

Lyons AC, Coronado AE, Piller F, Poler R (2012) Network collaboration: vertical and horizontal partnerships. In: Lyons AC, Coronado AE (eds) Customer-driven supply chains. Springer, London, pp 95–111

Mason R, Lalwani C, Boughton R (2007) Combining vertical and horizontal collaboration for transport optimization. Supply Chain Manag Int J 12(3):187–199

Sáenz MJ, Revilla E, Knoppen D (2014) Absorptive capacity in buyer–supplier relationships: empirical evidence of its mediating role. J Supply Chain Manag 50(2):18–40

World Economic Forum (2009) Supply chain decarbonisation. The role of logistics and transport in reducing supply chain carbon emissions. Geneva

Zaragoza Logistics Center. www.zlc.edu.es

Chapter 2
Inter-organizational Learning and Collaboration

Companies are increasingly aware that combining critical resources with strategic partners, such as supply chain partners, may provide competitive advantages (Paulraj et al. 2008). Supply chains are moving toward complex, collaborative value networks in which partners work and experiment together on problem solving, promoting inter-firm learning, and sharing risks and benefits (Malhotra et al. 2005). The value of this success is reflected in how firms like Zara, Procter & Gamble, Toyota, and Dell have used their collaborative relationships as competitive weapons to gain advantages over competitors (Dyer and Singh 1998). Nonetheless, despite the existence of some successful cases, practice shows the difficulty of strategic collaborative relationships for many reasons, including lack of trust between trading partners (Johnston et al. 2004) and a lack of alignment between the mental models of the partners involved (Sterman 2000). One of the most fruitful ways of addressing these weaknesses is through inter-organizational learning practices.

The learning process underlying collaborative relationships is both inter-organizational and intra-organizational (Vera and Crossan 2006), but in practice it is difficult to connect these two facets (Fawcett et al. 2011). Boundary spanners frequently obtain novel ideas and insights from partner companies but encounter difficulties in effectively applying this knowledge in their own firms (Hult et al. 2000).

Insights into the connection between inter- and intra-organizational learning processes remain limited, although the absorptive capacity construct (AC) provides a suitable starting point for exploring this connection (Cohen and Levinthal 1990; Easterby-Smith et al. 2008; Holmqvist 2004). Absorptive capacity has been widely studied since its inception by Cohen and Levinthal (1990), among others, who defined the construct as "the ability of a firm to recognize the value of new, external information, assimilate it, and apply it to commercial ends," (1990, p. 128). Absorptive capacity has been used in more than 1,000 peer-reviewed

M.J. Saenz et al., *Enabling Horizontal Collaboration Through Continuous Relational Learning*, SpringerBriefs in Operations Research,
DOI 10.1007/978-3-319-08093-2_2

academic articles since 1990. Although originally developed in an R&D context, AC has potential explanatory power in a collaborative-relationship context given that key suppliers or customers constitute important sources of new knowledge (Hult et al. 2007; Sáenz et al. 2014), and considering the increased efforts spent coordinating and integrating business activities between buying and supplying firms (Krause et al. 2007). When AC is discussed in a relational context, the major source of new knowledge comes from a collaborative partner, and according to its original definition it is represented by the three complementary processes mentioned above: exploration, assimilation, and exploitation (Lane et al. 2006).

References

Cohen WM, Levinthal DA (1990) Absorptive capacity: a new perspective on learning and innovation. Adm Sci Q 35:128–152

Dyer JH, Singh H (1998) The relational view: cooperative strategy and sources of interorganizational competitive advantage. Acad Manag Rev 23(4):660–679

Easterby-Smith GM, Antonacopoulou E, Ferdinand J (2008) Absorptive capacity: a process perspective. Manage Learn 39(5):483–501

Fawcett SE, Wallin C, Allred C, Fawcett AM, Magman GM (2011) Information technology as an enabler of supply chain collaboration: a dynamic-capabilities perspective. Journal of Supply Chain Manag 47(1):38–59

Holmqvist M (2004) Experiential learning processes of exploitation and exploration within and between organizations: an empirical study of product development. Organ Sci 15(1):70–81

Hult GTM, Hurley RF, Giunipero LC, Nichols EL Jr (2000) Organizational learning in global purchasing: a model and test of internal users and corporate buyers. Decis Sci 31(2):293–325

Hult GTM, Ketchen DJ, Arrfelt M (2007) Strategic supply chain management: improving performance through a culture of competitiveness and knowledge management. Strateg Manag J 28(10):1035–1052

Johnston DA, McCutcheon DM, Stuart FI, Kerwood H (2004) Effects of supplier trust on performance of cooperative supplier relationships. J Oper Manage 22:23–38

Krause DR, Handfield RB, Tyler BB (2007) The relationship between supplier development, commitment, social capital accumulation and performance improvement. J Oper Manage 25:528–545

Lane PJ, Koka BR, Pathak S (2006) The reification of absorptive capacity: a critical review and rejuvenation of the construct. Acad Manag Rev 31(4):833–863

Malhotra A, Gosain S, El Sawy OA (2005) Absorptive capacity configurations in supply chains. MIS Q 29(1):145–187

Paulraj A, Lado AA, Chen IJ (2008) Inter-organizational communication as a relational competency: antecedents and performance outcomes in collaborative buyer-supplier relationships. J Oper Manage 26:45–64

Sáenz MJ, Revilla E, Knoppen D (2014) Absorptive capacity in buyer-supplier relationships: empirical evidence of its mediating role. J Supply Chain Manage 50(2):18–40

Sterman J (2000) Business Dynamics. McGraw-Hill, Irwin

Vera D, Crossan MM (2006) Organizational learning and knowledge management: towards an integrative framework. In: Easterby-Smith M, Lyles MA (eds) Handbook of organizational learning and knowledge management. Wiley-Blackwell, Oxford, pp 122–141

Chapter 3
Vertical Collaboration and Horizontal Collaboration in Supply Chain

Collaboration in supply chains has been defined as occurring when "two or more independent companies work jointly to plan and execute supply chain operations with greater success than when acting in isolation" (Simatupang and Sridharan 2002). It can occur in many ways and is commonly divided into two main categories (see Fig. 3.1): (1) vertical, when collaborating with customers, internally (across functions), service providers and with suppliers; and (2) horizontal, between different supply chains when cooperating with competitors and with non-competitors (Barrat 2004).

The term "supply chain management" itself describes vertical collaboration (Cruijssen et al. 2007) because it captures all of the approaches used by the different suppliers, manufacturers, and distribution centres to successfully integrate their operations and create an efficient supply chain (Simchi-Levi et al. 2000). It is, thus, aimed at providing all the necessary links and cooperation practices to have a seamless production and distribution network.

Collaborative practices can definitively facilitate the connection of all network operations among the different nodes. Collaborative planning, forecasting, and replenishment (CPFR) emerged in the mid-1990s as the first kind of collaboration in the context of supply chains (Barrat 2004). Further examples of vertical cooperation are efficient customer response (ECR), and vendor-managed inventory (VMI) (Cruijssen et al. 2007). The former emerged also in the 1990s as an initiative of the grocery and consumer packaged goods (CPG) industries, and it changed the managerial philosophy from holding information internally to sharing it in order to improve customer value (Simatupang and Sridharan 2002). It served as the launch for vendor-managed inventory (VMI) and continuous replenishment (CR) (Whipple and Russell 2007). Nevertheless, the scope of collaboration in supply chains did not stand at a vertical dimension (Barrat 2004). Companies also started to look into other supply chains in order to find beneficial partnerships. Although some horizontal collaboration examples can be found, like the Manufacturers Consolidation Centers (MCCs) (Cruijssen et al. 2007), this kind of cooperation in supply chain remains still relatively embryonic in its current development.

M.J. Saenz et al., *Enabling Horizontal Collaboration Through Continuous Relational Learning*, SpringerBriefs in Operations Research,
DOI 10.1007/978-3-319-08093-2_3

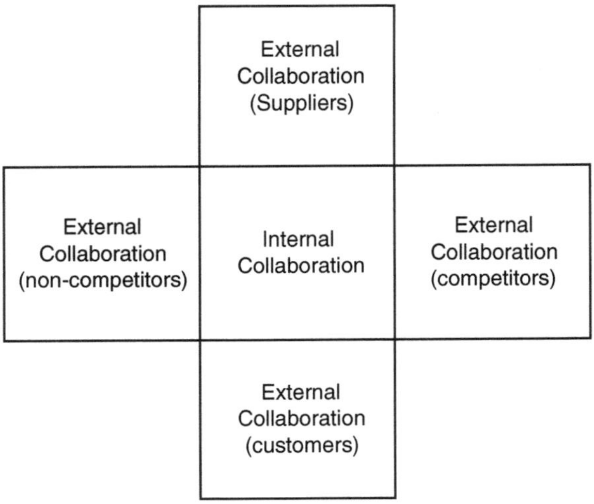

Fig. 3.1 Forms of collaboration (adapted from Barrat 2004)

In this sense, when looking for opportunities beyond the company's value chain, transport has been considered a versatile asset for supply chain improvement, and horizontal collaboration has proved to be an important element in distribution optimization since it exploits better the conceptualization of supply chain as supply networks (Mason et al. 2007). Cruijssen et al. (2007) present an initiative of eight Dutch producers of sweets and candies that intensively cooperated by supplying 250 drop-off points and contracting with a logistics service provider that would consolidate the shipments and achieve delivery efficiency (Cruijssen et al. 2007). In a later paper, Cruijssen et al. (2010) evaluate the insights of horizontal collaboration from the perspective of 82 road transportation companies in Flanders. One of the main conclusions is that horizontal collaboration is not easy to implement due to the apprehension shown by non-cooperating companies, which "tend to underestimate the opportunities and overestimate the impediments" of horizontal collaboration (Cruijssen et al. 2010: 167). Impediments are barriers perceived by the companies, which make them reluctant to start collaborating. These barriers include a lack of trust and the fear of sharing sensitive information (Everinghton et al. 2010). The opportunities are the driving forces (Cruijssen et al. 2007) behind the start-up of alliances. Examples of such motivations are reducing transport cost and accessing new markets (Lyons et al. 2012). Companies willing to collaborate have the expectation of getting benefits out of their business relationships (Parkhe 1991). Following De Clercq and Sapienza (2001), collaboration can develop a combination of firm-specific benefits through inter-firm commitment that creates relational rents. These can be characterized as "supernormal profit jointly generated in an exchange relationship that cannot be generated by either firm in isolation and can only be created though the joint idiosyncratic contributions of the specific alliance partners" (Dyer and Sing 1998; Sáenz et al. 2014).

The model proposed in this book intends to serve as a guide for overcoming the barriers to horizontal collaboration and for exploiting the drivers in order to create specific relational rents. It gathers the main elements that enhance the continuous relational learning that helps to create absorptive capacity that will generate relational rents. The relational view (Dyer and Singh 1998) asserts that trusting one another (Lado et al. 2008) takes the cooperating parties further in generating value than simply relying on contracts (Branstad 2006). Sometimes, in combination with formal controls (Granovetter 1985), there are further informal safeguards that firms rely on, such as solidarity, mutuality, and dependence, that enable collaboration to be effective and sustainable. The long-term maintenance of a relationship builds solidarity between partners (Heide and John 1992), and the behavioural expectations are based on the mutuality of the parties' interests (Bello et al. 2003). The relational rents will be the product of the complementarity agreement of foregoing one's interests at the expense of others (Powell et al. 1996). This will lead to a dependence of the parties on one another (Pfeffer 1997) and will serve as a deterrent against opportunism (Wilding and Humphries 2006). The empirical research used to develop this model is based on embedded case studies (Yin 2003) and claims to prove how the former constructs enhance horizontal collaboration and the creation of absorptive capacity to gain relational rents, while evaluating the main impediments to and drivers of this kind of cooperation.

References

Barrat M (2004) Understanding the meaning of collaboration in the supply chain. Supply Chain Manage: Int J 9(1):30–42

Bello DC, Cheraliu C, Zhang Li (2003) The antecedents and performance consequences of relationalism in export distribution channels. J Bus Res 56:1–16

Branstad A (2006) Learning by grafting knowledge—report from an ongoing case study in an innovation centre. 14th Nordic Conference on Small business Research, May 11–13, 2006

Cruijssen F, Dullaert W, Fleuren H (2007) Horizontal cooperation in transport and logistics: a literature review. Transp J 46(3):22–39

Cruijssen F, Dullaert W, Joro T (2010) Freight transportation efficiency through horizontal cooperation in Flanders. Int J logistics: Res Appl 13(3):161–178

De Clercq D, Sapienza HJ (2001) The creation of relational rents in venture capitalist-entrepreneur dyads. Venture Capital 3(2):107–127

Dyer JH, Singh H (1998) The relational view: Cooperative strategy and sources of interorganizational competitive advantage. Acad Manag Rev 23(4):660–679

Everington L, Lyons A, Li D (2010) Horizontal collaboration in the logistics industry. In: Whitheing T (ed) Proceedings of the LRN conference 2010: towards the sustainable supply chain, CILT (UK), Harrogate, 204–211

Granovetter M (1985) Economic action and social structure: the problem of embeddedness. Am J Sociol 91(3):481–510

Heide JB, John G (1992) Do norms matter in marketing relationships? J Mark 56(2):32–44

Lado AA, Dant RR, Tekleab AG (2008) Trust-opportunism paradox, relationalism and performance in interfirm relationships: evidence from the retail industry. Strateg Manag J 29:401–423

Lyons AC, Coronado AE, Piller F, Poler R (2012) Network collaboration: vertical and horizontal partnerships. In: Lyons AC, Coronado AE (eds) Customer-driven supply chains. Springer, London, pp 95–111
Mason R, Lalwani C, Boughton R (2007) Combining vertical and horizontal collaboration for transport optimization. Supply Chain Manage: Int J 12(3):187–199
Parkhe A (1991) Interfirm diversity, organizational learning and longevity in global strategic alliances. J Int Bus Stud 20:579–601
Pfeffer J (1997) New directions for organization theory: problems and prospects. Oxford University Press, New York
Powell WW, Koput KW, Smith-Doerr L (1996) Inter-organizational Collaboration and the Locus of Innovation: networks of Learning in Biotechnology. Adm Sci Q 15(1):88–102
Sáenz MJ, Revilla E, Knoppen D (2014) Absorptive capacity in buyer-supplier relationships: empirical evidence of its mediating role. J Supply Chain Manage 50(2):18–40
Simatupang TM, Sridharan R (2002) The collaborative supply chain. Int J Logistics Manage 13(1):15–30
Simchi-Levi D, Kaminsky P, Simchi-Levi E (2000) Designing and managing the supply chain: concepts, strategies, and cases. McGraw-Hill, New York
Whipple JM, Russell D (2007) Building supply chain collaboration: a typology of collaborative approaches. The international Journal of Logistics Management 18(2):174-196
Wilding R, Humphries AS (2006) Understanding collaborative supply chain relationships through the application of the Williamson organisational failure framework. Int J Phys Distrib Logistics Manage 36(4):309–339
Yin RK (2003) Case study research. design and methods. Sage Publications Inc, Thousand Oaks

Chapter 4
Model for Horizontal Collaboration Dynamics

When they cooperate, companies expect to take advantage of the knowledge transfer that occurs with the information exchange. In order to obtain such advantages, they have to recognize the value underlying their interactions, assimilate it, and apply it to commercial ends. The ability of a firm to do so has been described in the previous chapter as "absorptive capacity." The creation of this capability is not expected to be absolute but rather relative to the relationship (Lane and Lubatkin 1998; Lane et al. 2001). It is, thus, an ability dependent on the type of collaboration and the settlement of the elements that influence the health of the relationship. Therefore, the key question when developing a generic model for horizontal collaboration is how to particularise the creation of absorptive capacity (Lane et al. 2006) for this specific kind of cooperation. This would imply building the relationship on a learning dynamic that supports the exchange of knowledge and the three phases of absorptive capacity creation: exploration, assimilation, and exploitation. These concepts have proved to be useful when applied to the supply chain management context (Sáenz et al. 2014). The model described in this chapter is the final result of an iterations-development process used throughout the case studies that are presented in Chap. 5. Starting from an initial framework built upon the idea of creating relational rents out of detected opportunities through relational learning, the case studies have highlighted all of the insights that enable or might disable the creation of absorptive capacity as well as the outcomes expected from the alliances. The different experiences of the analysed companies and the maturity of their collaborative practices have allowed us to organize these influencing elements in a structured way to serve as a guide for starting and developing profitable and sustainable horizontal collaborations. In Chap. 6, the reader will find a summary of how the model has been applied in two successful horizontal collaboration consortiums.

M.J. Saenz et al., *Enabling Horizontal Collaboration Through Continuous Relational Learning*, SpringerBriefs in Operations Research,
DOI 10.1007/978-3-319-08093-2_4

4.1 Drivers and Barriers

The sector and location of a company determines its competitive frame, its legal position, and its available infrastructure within a country—i.e., its environment—according to which a company will find different drivers/barriers that influence collaborative attitudes. For the purpose of this research, a driver is any kind of stimulation coming from the environment or from inside the organization that encourages a company to find a partner with which to start horizontal collaboration. A barrier is any kind of obstacle that creates negative thinking about collaboration as a value-added practice or about establishing cooperation once the company has decided to search for a partner.

The proposed model is built upon the idea that these two constructs will be either the starting point for collaboration in the case of a driver or a "no-go" element disabling collaborative initiatives, which are drawn on the left side of Fig. 4.1. The model also considers that the environment is in constant change, which means that new opportunities and impediments will appear over time. For this reason, and because we are seeking to be as generic as possible, the model is not limited to a given list of specific drivers or barriers.

Nevertheless, as an overview of real examples that have affected companies from diverse sectors and different maturity levels in collaboration practices, the reader will find in Chap. 6 a summary of the main opportunities and impediments assessed by the interviewed managers in the case studies and by the companies involved in the consortiums.

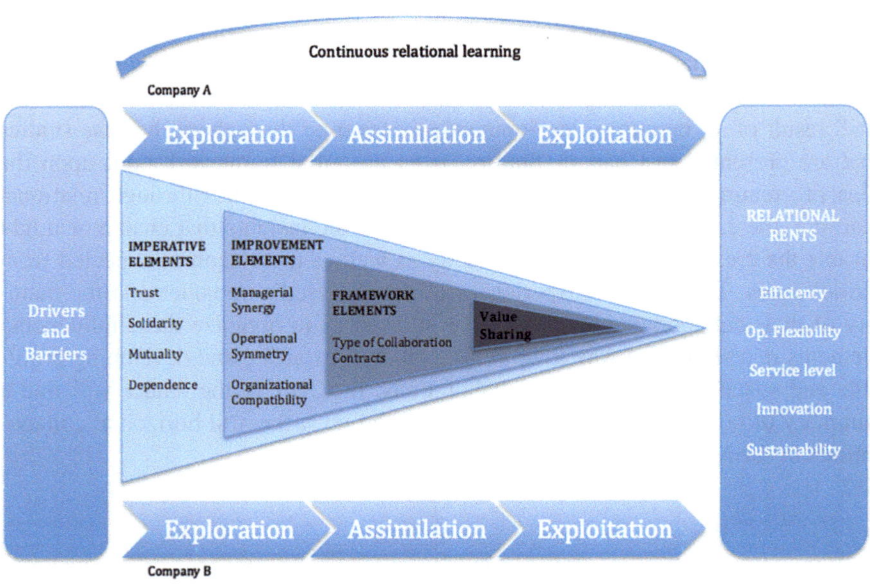

Fig. 4.1 Horizontal collaboration model

4.2 Relational Rents

When starting an alliance, the companies expect a profit out of the relation that they would not be able to generate on their own. As mentioned earlier, these supernormal benefits are called relational rents and, as outcomes of the partnership, are placed on the right side of the model. In order to be specific for the field, the model had to capture the most relevant expected outcomes after implementing horizontal collaboration. Gartner (2013) has shown that improvement of the service level, innovation, efficiency, and sustainability are among the top supply chain business priorities.

Collaborating horizontally with companies at the same level but from different supply chains is a perfect vehicle for achieving efficiency levels they could never reach when operating in isolation. Increasing the efficiency of freight distribution implies reduced costs, and this can be achieved by economies of scale derived from the common shipments. Service frequency is a major concern given the increasingly demanding expectations of customers and the actual economic costs of delivering. The pressure from customers for more frequent and faster deliveries of smaller quantities forces companies to continuously search for new ways to improve their service levels without incurring higher costs. For companies with small delivery volumes, cooperating with a company with a similar delivery profile would produce cost- and service-level efficiencies. This would also enable greater flexibility in meeting customer demands. However, it may also be an advantage to have partners in collaboration with dissimilar service levels. An example might be one company, A, that moves goods daily between two locations partnering with another company, B, that moves weekly between the same two locations. In this situation, while implementing joint horizontal collaboration practices, company B has the opportunity to improve its service level at virtually no cost, and company A has the flexibility to incorporate company B's product to suit and maximize the load volumes. Taking the environmental dimension of the term "sustainability," the reduction of CO_2 emissions is a rent derived from the improvement of the load factor. Environmental aspects of supply chain management have been leading factors for sustainable supply chains (Carter and Easton 2011), and the latter is an outcome the companies expect from collaboration since corporate social responsibility and emission restrictions are pushing them to find solutions for reducing their carbon footprint.

The global marketplace has also driven companies to innovate. In this sense, horizontal collaboration is a source of innovation through inter-organizational learning. Cooperating with other firms that operate at the same level in their supply chains is an ideal framework for joint innovation since common risks and opportunities are shared.

The top priorities assessed by Gartner, together with operational flexibility, are the relational rents considered as outcomes in this model of horizontal collaboration, which will be challenged by the business cases depicted in Chap. 5.

4.3 Learning Process Dynamics

The purpose of this model is to enable dynamics for continuous improvement of the supply chains of both companies in a collaboration. The benefits of a horizontal collaboration are considerable, but they are only reachable and sustainable in the long run if the companies in the collaborative partnership are willing to work on developing a continuous relational learning environment. This means exploring together the potential improvements of both supply chains, assimilating each one internally within their own organizations, making the changes they have to make in order to enable the collaboration, and exploiting these changes through joint management and execution of the collaborative practices.

These dynamics will be the way to continuously strengthen the relationship and make the cooperation more efficient by developing the elements described in this section, and they will generate outcomes or relational rents out of the opportunities or drivers.

These dynamics can be better understood with an example, as follows.

Given partners A and B from different sectors, which share, for instance, a common low-demand area, they will face the opportunity to collaborate in order to improve the delivery frequency to their customers. Their driver for collaboration would be to send two half trucks each 2 days, better than one full truck each 4 days or one truck half loaded every 2 days.

In the exploration phase they should first evaluate the partner's economic and corporate performance, as well as other organizational practices, to find out if they can trust the partner. They should verify whether the two companies' business philosophies, objectives, and corporate cultures are well aligned. This would imply the interchange of each company's "values portfolio." Given an initial positive assessment, they should then study their potential synergies, which means identifying the ways in which the dyad can be more efficient than the two firms individually. Analysing their demand patterns, they should determine when and over which routes they can collaborate with the partner to be potentially more efficient, to fully load their trucks, and to improve their flexibility in satisfying their customers. They should also study the two firms' compatibility in terms of the degree of alignment of their resources, such as handling and transportation equipment and technology, among other things. In this case, they should assess whether their distribution nodes are close enough and if their packing and delivery specifications are compatible. Once this operational plan has been designed, both companies will have to get into the assimilation phase, in which they train their employees for the collaboration in order to move from the top down into the inter-organizational teams that are going to work together. They will also adapt their current procedures, processes, and information systems to the new horizontal collaboration requirements. Each of the two organizations will do these things internally, while also taking into consideration the fact that the goal is for the two companies to work together.

Finally, they will identify the optimal type of collaboration and contract according to their needs. This should produce a well-structured system of sharing gains of the collaboration outcomes.

In the exploitation phase, the companies will start with specific collaborative actions. After exploring the processes they have to change and assimilating those changes internally, they can finally align and share the transportation flows and profits from the advantages the collaboration offers in terms of improved frequency and service levels.

These dynamics are placed in a continuous-improvement loop due to the dynamics of both firms' businesses that allows relational learning to take place, with the outputs of the exploitation phase being the starting points for further exploration. Following the example, after exploiting this opportunity, once it has proved to be a profitable practice, the companies, seeking to establish a more regular and sustainable cooperation, should enter again into the exploration phase in order to best align elements that would enhance a more regular collaboration, like setting a more robust operational plan, some additional contractual agreements, etc. Then, again, they will have to assimilate the new changes and exploit the opportunities through joint execution, hopefully in a more structured and improved way, to achieve better outcomes from the horizontal-collaboration practices.

This is how the loop of relational learning would work and how it would continuously help the dyad to become more mature, stronger, and more profitable. The two partners should always be aware of the possible upturns of relational learning and the improvement potential they offer to their alliance.

4.4 Elements of the Horizontal Collaboration Dynamics Model

The elements displayed in the centre of Fig. 4.1 are the aspects that should be addressed in order to best create absorptive capacity to obtain relational rents out of the collaboration and to make it sustainable over the long run. They are organized in a hierarchical way, with the "imperative elements" at the bottom of the pyramid, moving up through the "improvement elements" to the "framework elements" and finally to value sharing. It means that, although collaboration can occur without having everything settled, it is expected that, the more mature the relationship is, the higher will be the level of the elements that can be deployed, which will make the collaboration more fruitful and sustainable by constantly creating absorptive capacity and exploiting it to increase the relational rents. It does not mean that, when a certain maturity in the collaboration has been achieved, the elements at lower levels can be forgotten. Rather, it means that the parties should not consider putting efforts into higher-level elements without having accomplished a certain level of joint performance on the lower-level elements. Once the partners have begun to assess the new elements, the learning processes should be encouraged to continue for all lower-level elements as well, since strengthening them will drive the collaboration toward more sustainable and profitable outcomes.

4.4.1 Imperative Elements

These are the elements that enable collaboration to develop from a one-time experience into regular, ongoing cooperation. A company needs to positively assess them with regard its potential partner in order to judge whether it reliable for collaboration. Adopting the definition of trust as one party's "confidence in the goodwill" of an exchange partner (Ring and Van de Ven 1994, p. 93), the companies will never cooperate with a partner that they cannot completely trust or that does not have a good reputation. Another major concern with regard to long-term collaboration is the element of solidarity, which deals with the fear of being left behind when facing difficult situations. Solidarity is defined as "a bilateral expectation that a high value is placed on the relationship," and "it prescribes behaviours directed specifically toward relationship maintenance" (Heide and John 1992, p. 26). As long as solidarity exists, the problems that arise can be solved as a team, and both companies will work together to achieve the common objectives. Another typical issue in all kinds of business relationships is the question: What if it is always me who contributes the greatest effort to solve problems? This topic, which is common in all types of collaboration, has been treated in the literature under the term "mutuality." In other words, "Even if the efforts, costs, and benefits are not always evenly shared, they balance out in the long term" (Lado et al. 2008). This mutuality between partners has its roots in interdependence and is essential to establishing a successful collaboration (Thomson et al. 2007). When companies identify unique resources from the partners' side that will guide them to mutual benefits, they feel the dependence of jointly working with this specific partner to achieve the relational rents. As long as the partners, which are dependent upon one another's resources, can satisfy their interests without hurting themselves, collaboration can occur (Wood and Gray 1991). The only way to judge companies' trust, solidarity, and mutuality is by analysing their business attitudes, and this is essential before any discussions about collaboration are begun. Usually, these features in a relationship are detected from previous experience or reputation.

4.4.2 Improvement Elements

Improvement elements are the things that, when mutually improved, enhance already-existing, regular, short-term collaborative practices. In this sense, three elements have been pointed out in the interviews: managerial synergy, operational symmetry, and organizational compatibility. It is, for example, important that a single company does not assume overall managerial control of a collaboration. Managerial synergy means that joint management is better than each company's individual control because it enables the partners to achieve outcomes that would be unobtainable if they were acting independently. Synergy can also come from coordination of logistical planning and operational processes. Being "operationally symmetric" is essential in order for a collaboration to be successful because the profiles of the partners' logistical operations have to coincide in terms of type of transport, time windows, locations visited, and distribution networks. With the right symmetry, companies can benefit from improved capacity utilization and reduced empty running.

For example, having similar handling and transportation technology, or operating within the same logistics cluster with facilities close together, can enhance capacity utilization. Facilities that are dispersed but within each other's delivery areas can reduce empty running by back hauling loads. The partners will also analyze their organizational compatibility. Past research has shown that organizational compatibility influences absorptive capacity, which in turn shapes relational benefits (Sáenz et al. 2014). Firms will be better able to learn from and collaborate with one another when the organizations have compatible norms and values, which highlights the role of cultural compatibility as a relevant antecedent for making sense of external knowledge and maximizing capability development. In this sense, the partners will assess the degree of alignment of goals and objectives as well as business philosophies and corporate cultures. From the operational point of view, they will explore whether their packaging and delivery specifications are compatible.

Ultimately, the improvement elements will be arranged throughout the learning process to make the collaboration as efficient as possible.

4.4.3 Framework Elements

Framework elements are the ones that set up the rules for long-term collaboration. Contracts are defined as the "parties' manifestations of mutual assent" (Slawson 1984). In this book we do not dig deeply into the formal aspects of a legal contract for horizontal collaboration (for that purpose, refer to the recommendations from the CO^3 European project, which are available at www.co3-project.eu). However, although "the effective implementation of normative contracts relies on the establishment and use of behavioural rules" (Lado et al. 2008) such as the imperative elements, as horizontal collaboration develops and grows, formal or informal contracts will become a necessity. In that regard, for example, given a certain level of trust between the partners, the best way to commonly explore how to achieve the best logistical efficiency levels could be the "open-books" type of contracts, in which the different transactions are transparent and then the gain sharing is easier to implement.

As previously mentioned, the model asserts that the companies are going to be able to make their collaboration more successful based on the absorptive capacity they are able to create, which is enhanced by the strengthening of all the elements. Once they have the will to engage in long-term cooperation, the type of collaboration is going to be a key element to be discussed.

As advanced by Lambert et al. (1999), there are three types of collaboration depending on the level of integration. Within the context of horizontal collaboration in the first stage of the relationship, where the companies are starting to know each other and might be still a bit reluctant to share sensitive information, they will focus on short-term cooperation, such as joint distribution or line hauling, back loading, purchasing/tendering groups, etc. In the second stage of the relationship, as the companies become more willing to share operational information, they will experiment with such things as synchronized planning, multi-modal collaboration, and warehouse/cross-dock sharing. At the last stage of the relationship, when the companies have gained substantial knowledge of each other through continuous relational learning, they are open to sharing sensitive information, which allows them to start "no-ending-date" collaboration, such as network integration or joint innovation.

4.4.4 Value Sharing

In discussions with the companies examined in this study, there are no definite rules for sharing the benefits when implementing horizontal collaboration. In reality, for simplicity, savings are usually randomly allocated, ranging from a 50/50 basis to a 70/30 basis. Where a logistics service provider is used, the shippers rely on it to set rates that are satisfactory to both parties. Some of the interviewees said that they would only enter a horizontal collaboration with competitors as long as there was a fair sharing mechanism according to the contribution of each partner. The equivalent cost reduction between partners that are non-competitors becomes a secondary issue as long as they can obtain the savings that they expected and as long as both partners contribute to achieving the desired outcomes, such as efficiency (cost savings), flexibility, service level, innovation, and sustainability. Game theory can help the partners to establish mechanisms to quantify the gain sharing in horizontal collaboration, through the application of the Shapley value (Cruijssen et al. 2007; Shapley 1953).

Lloyd S. Shapley, one of the first researchers in the area of game theory, formulated the problem of how to distribute the payoffs of cooperation. He suggested some properties that a fair distribution rule should fulfil:

1. Efficiency: in our case, that all of the savings which can be obtained are split evenly among the players.
2. Symmetry: the value must be anonymous (symmetric) in the sense that what one player wins is not a function of their labelling but rather of their contribution to the game.

3. Linearity: it seems reasonable to propose that what the players obtain should not depend on the distribution protocol, for example on whether one part is first distributed and then the rest.
4. Null player: whoever does not contribute should not harm the others (this person should not receive any compensation) or pay any penalties.

The main outcome of Shapley's theory (1953) is that there is a unique value which the axioms of efficiency, symmetry, linearity and null player should satisfy meaning there is one possible answer to any given distribution problem. The mathematical formula and theoretical demonstration can be consulted in Shapley (1953).

One important component of horizontal collaboration value sharing is how the members of the consortium are able to share the risks involved. A series of risks were identified throughout the interviews conducted in this research. The most often mentioned was a risk of worsening the service level due to delivery delays deriving from planning failures by a given partner. To this end, there may be a conflict of interest when companies are at similar service levels. Who should then have priority as far as complying with the time window? These are concerns that imply a commitment to which some firms would prefer associating written responsibilities.

Firms that currently collaborate asserted they were willing to sacrifice at times for the good of the relationship as long as the impacts were not too significant. Whether in writing or not, all of them showed a concern for managing and minimizing the inherent risks of HC.

References

Carter CR, Easton PL (2011) Sustainable supply chain management: evolution and future directions. Int J Phys Distrib Logis Manag 41(1):46–62

Cruijssen F, Dullaert W, Fleuren H (2007) Horizontal cooperation in transport and logistics: a literature review. Transp J 46(3):22–39

Gartner (2013) State of the industry: opening address. Retail and FMCG Supply Chain Summit, Eye for Transport, Dusseldorf

Heide JB, John G (1992) Do norms matter in marketing relationships? J Market 56(2):32–44

Lado AA, Dant RR, Tekleab AG (2008) Trust-opportunism paradox, relationalism and performance in interfirm relationships: evidence from the retail industry. Strateg Manag J 29:401–423

Lambert DM, Emmelhainz MA, Gardner TJ (1999) Building successful logistics partnerships. J Bus Logis 20(1):165–181

Lane PJ, Lubatkin M (1998) Relative absorptive capacity and interorganizational learning. Strateg Manag J 19(5):461–477

Lane PJ, Salk JE, Lyles MA (2001) Absorptive capacity, learning, and performance in international joint ventures. Strateg Manag J 22:1139–1161

Lane PJ, Koka BR, Pathak S (2006) The reification of absorptive capacity: a critical review and rejuvenation of the construct. Acad Manag Rev 31(4):833–863

Ring PS, Van de Ven AH (1994) Development processes of cooperative inter-organizational relationships. Acad Manag Rev 19:90–118

Sáenz MJ, Revilla E, Knoppen D (2014) Absorptive capacity in buyer–supplier relationships: empirical evidence of its mediating role. J Supply Chain Manag 50(2):18–40

Shapley LS (1953) A value for n-person games. In: Kuhn HW, Tucker AW (eds) Contributions to the theory of games, vol II. Princeton University Press. Ann Math Stud 28:307–317
Slawson WD (1984) The new meaning of contract: the transformation of contracts law by standard forms. University of Pittsburgh Law Review, Pittsburgh
Thomson AM, Perry JL, Miller TK (2007) Conceptualizing and measuring collaboration. Oxford University Press, Oxford
Wood DJ, Gray B (1991) Toward a comprehensive theory of collaboration. J Appl Behav Sci 27(2):139–162

Chapter 5
Case Studies

5.1 Company Needs

5.1.1 Methodology

Due to the lack of an empirical analysis in this field, the research methodology of this book had to be built upon an exploratory, multi-case study (Yin 2003). It contains interviews with five senior managers from three different industries: home appliances, automotive manufacturing, and retail sales. We complement the empirical research with evidence collected on the horizontal collaborative successful practices the firms have implemented.

In order to develop the model in such a way that it can serve as a universal guide for horizontal collaboration, the companies were selected based on their diversity. They operate in different regions, with different infrastructure availabilities, and they are at different echelons of the supply chain. Two of the companies are manufacturers and the other two operate in the retail business. All of the interviewed managers had strong experience in logistics and supply chain management, as well as senior responsibility for the supply chain, including procurement and logistical strategies. They play roles as boundary spanners for both potential horizontal collaboration and already-existing distribution cooperation. Their positions at the interface between their firms and the partner companies allow them to contribute to the collaboration process based on their long experience (Saenz et al. 2014). Their companies are also at different maturity levels in terms of horizontal collaboration development, and this gave us the opportunity to validate the proposed model for any given maturity grade (Table 5.1).

Semi-structured interviews were determined to be the best way to obtain the viewpoints of the managers about the set of previously selected discussion issues in the model presented in the previous section. This approach did not impede the interviewees from speaking freely about their experiences and, thus, providing us

M.J. Saenz et al., *Enabling Horizontal Collaboration Through Continuous Relational Learning*, SpringerBriefs in Operations Research,
DOI 10.1007/978-3-319-08093-2_5

Table 5.1 Interviewed companies and interviewees

Company	1	2	3	4	5
Sector	Home appliances	Automotive	Home-improvement and gardening	Hypermarket	Hypermarket
Number of employees	4,000	7,000	8,000	50,000	50,000
Supply chain role	Manufacturer	Manufacturer	Retailer	Retailer	Retailer
Position	Logistics director	Head of logistics	Director of information systems and supply chain	Procurement director	Development and projects director

with insights about the stages of collaboration they were facing. The interview protocol (Annex I) contains 38 questions. Four of them were specific questions about their company (general information, sector, company size, etc.), and 13 of them were Likert-scale questions asked with the purpose of ranking the drivers and barriers, varying from 1 (does not encourage or discourage companies from starting a collaboration), up to 5, (strongly encourages or discourages horizontal collaboration). There were additional open questions about their company's culture, their general collaborative practices, their specific experiences with horizontal collaboration, the way they started, and how it became a regular cooperation or why it ended. Following their assessments, we were able to develop the proposed model and implement it in the two successful horizontal collaboration consortiums that are presented within this chapter.

5.1.2 Home Appliances Manufacturer

The first interviewee was the Managing Director of the Logistics Department in a multinational home-appliances manufacturer. The company employs approximately 4,000 people in its home country and about 40,000 worldwide. It has seven factories in-country, and it exports products to 40 different countries. Facing over 50 competitors in its product lines, it is the leader in this complex market.

Within their company culture, values such as sustainability and social responsibility have driven them over the past few years to collaborate in different areas of the business, such as with universities in R&D. Specifically in supply chain, they have developed internal horizontal collaboration between their factories, by commonly planning the international distribution needs of half-finished and finished goods between the production plants and warehouses, and external vertical collaboration, by insourcing activities to many in-house providers. The company has always been open to engaging in collaboration, and it is experienced in collaboration processes. Because of its long experience, the company is careful to avoid linking its name to another company with bad commercial practices.

When asked about external horizontal collaboration, the manager described evidence of collaboration with another big company from the mattress industry. Both have warehouses in the same city, from which they distribute their products to the entire country. One particular area they both serve has very low demand rates. The main driver for this collaboration was the desire to achieve easier and quicker response to demand fluctuations. After having conducted an investigation of the partner company's values and financial health in order to assess trust and organizational compatibility, the manager's company established operational symmetry of their distribution routes and started to increase their delivery frequency by filling the trucks together with the new partner. For example, while formerly they had to either send a half-full truck every two days, or deliver with a full truck every four days, now the firm, by collaborating with the other company, has more flexibility in dealing with demand variations since they can fill a truck together every two days thus have provide improved service levels.

This collaboration, which started with informal conversations at a logistics conference, is built upon a verbal compromise to both push their common logistics service provider in order to exploit the opportunities for joint regional distribution. The relationship is in a developed but still ad hoc phase, with no formal contracts and no defined type of collaboration. Through many iterations of their joint distribution, they have been able to strengthen the "imperative elements" such as the mutuality of a common logistics provider, by recognizing that the benefits of the collaboration are equally shared over the long run, even though at certain points one of the partners might be profiting more than the other from the joint operations. At the same time, the "improvement element" called operational symmetry evolved over time from a basic alignment of the transportation routes into a joint exploration of the most efficient time windows for both companies, assimilation towards internal planning and exploitation of the agreed terms. Although the interviewee asserted that there was no need for binding agreements or defined collaboration types due to the mediation of the logistics provider, he admitted that if they would want to establish a joint, regular distribution plan, they would have to examine both the contractual terms and the collaboration framework, as well as negotiate a fair gain-sharing structure.

This collaboration is in an intermediate phase with regard to the degree of maturity in the collaboration because the companies have already taken advantage of information exchange. Nevertheless, the interviewee accepted the fact that there is still room for improvement. He wished to establish more regular cooperation through a formerly settled type of collaboration and a contract. This would build an even more reliable working atmosphere for knowledge transfer that would effectively help to create absorptive capacity and allow the partners to further explore opportunities and potential benefits from the alliance.

5.1.3 Automotive Manufacturer

The second interviewee was the Manager of the Logistics Department of a multinational automotive parts manufacturer. The company has 7,000 employees, and the industrial park in which the factory is located connects 63 suppliers with it, employing in total around 10,000 workers. As in the previous case, this multinational company has the advantage of being able to fill its trucks to an extent by efficiently planning the distribution of different half-finished products to its factories. Also, a very developed example of vertical collaboration exists in the in terms of the synchronized material flow from the suppliers around the factory directly to the assembly line via an overhead conveyor installed in a tunnel.

It was clear that the interviewee was not reluctant to collaborate with other firms since he had had a good experience in vertical collaboration with the providers in the industrial park. Also, horizontal collaboration in the distribution of finished vehicles is not a new practice since cars from different manufacturers are being jointly transported today. When the interviewee was asked about other sorts of horizontal collaboration, even with companies from other sectors, he offered a very interesting example. It was during a time of a large transport strike that he learned how horizontal collaboration could help solve a critical situation. Worried about the arrival of some supplies that may have had a huge impact on their production plans, the company found a firm from the milk industry that was located in the region from which the supplies were supposed to come, and that company offered to load the empty space in their finished-goods distribution trucks with the supplies needed by the manufacturer. Again, in a very informal way, an industrial manufacturer found an opportunity to solve a problem caused by an external contingency through horizontal collaboration. This synergy in their transportation management offered him an embedded solution to overcome an operational constraint. The collaborative practice is in an early stage since no analysis of the improvement elements has yet been done, but receiving help from another company in a difficult situation is already good evidence of solidarity, and the interviewee is, thus, now willing to jointly explore how this one-time cooperation could become standard practice in order to improve their operational flexibility.

There are other horizontal collaboration experiences in the automobile industry. Original-equipment manufacturers (OEMs) have shared railcars for outbound vehicle shipping, and there have been several major collaborations between manufacturers for the production and distribution of vehicles and between manufacturers and direct suppliers. In another example, a manufacturer and a supplier shared a distribution centre and inbound logistics for components.

New synergies in common horizontal logistical flows in the automotive sector can be obtained through supply chain integration, which in turn can result in a reduction in the number of delivery vehicles (including empty vehicles) and CO_2 emissions while also providing logistical cost savings. There are new key roles that need to be implemented in order to guarantee neutrality as well as fairness, and these roles can be played by professional associations and neutral institutions in the automotive industry.

5.1.4 Retailer A

The third interviewee works for a do-it-yourself retail chain that is also a multinational company, with approximately 8,000 employees in its home country. The interviewee is the Managing Director of Information Systems and Supply Chain. The retail chain manages more than 35,000 different articles in its commercial areas of between 6,000 and 14,000 m^2 and in its smaller stores all over the home country. The distribution is managed through a central hub in the home country and three cross-docking platforms. Their supplier network comprises approximately 900 suppliers, which include both small and large companies.

The interviewee had strong experience in horizontal collaboration and had experienced both good and bad results from this kind of cooperation. Despite some bad experiences, however, he says that he constantly looks for collaborative opportunities. At one point, the company was faced with a difficult situation with a logistics provider in the material flow from the suppliers to the central hub. The retailer suggested that the logistics provider should consolidate the loads of all their suppliers and guarantee a regular delivery on three different already-agreed-upon time fences over the week. In return, the retailer's multinational group would certify the excellence of this logistics service provider with a quality stamp, which would be expected to increase their turnover. The service provider did increase its income by threefold, but the agreed-upon terms were not upheld. The interviewee blamed it on a lack of internal assimilation by the logistics service provider, which was not able to implement regionally what they had agreed on at national level.

A horizontal collaboration practice that resulted in beneficial outcomes, though only on a few occasions, was a back-hauling exercise practiced with a manufacturer from another sector that had factories in a place where the retailer had some commercial areas. The idea was to fill the trucks on the way back from the stores with materials from the manufacturer in order to avoid an empty return. It worked, but it did not develop into a regular cooperation due to the lack of a joint planning platform. The distribution implemented ad hoc planning actions through telephone calls, which caused the interviewee's company to fear the risk of not coordinating efficiently and decreasing its service levels. In this case, the collaboration did not stop due to a bad assimilation but rather because they did not explore further elements, like managerial synergy.

After these two experiences, which were not entirely satisfactory, the interviewee did not give up and searched for an ideal partner. He found it in the food retail sector—a national company very well known for its operational dynamics and supply chain management expertise—with which the interviewee tried the same sort of arrangement that they had had with the manufacturer. In this case, the logistics service provider of the food retail chain would do the transport but both companies would have a platform for joint planning, and everybody would obtain benefits from the partnership. The companies would decrease their individual transport costs, and the logistics service provider would increase their total volume.

The outcome that the do-it-yourself retailer was searching for had an influence not only on the cost reduction but also on the joint innovation and the relational learning since they would profit from the food retail company's national distribution expertise and knowledge and would share in return their know-how on the international market, which their partner was willing to explore. This common benchmarking exercise based on common logistics of horizontal collaboration practice is expected to benefit the relational learning of the two retail companies.

The interviewee described their first meetings as a sharing of their companies' values, which turned out to be very compatible. After that, they explored and assimilated their transportation needs, including things as diverse as collection time windows and the kind of pallets that would be used. The manager said he did nothing new but only things he had tried in his former attempts at collaboration. However, the keys to establishing and maintaining the partnership in this case were (1) successful cooperation that was close to an integration of their planning through a common platform, and (2) the definition of a contract and additional value sharing via "open books." In other words, everybody knew how much the partner was paying and what the margin of the logistics provider was. Also, in case of an increase in the transportation cost for the logistics provider, it would be shared between the two partners.

This case study shows a very mature horizontal collaboration, with consideration of the elements and application of the explicit relational learning values so that everybody could benefit. They built a win-win relationship that made the cooperation sustainable over the long run.

5.1.5 Retailer B: Procurement

The fourth interviewee was the Procurement Director of a large supermarket retailer, which was different than the companies presented in the previous cases. The company has approximately 50,000 employees.

This interviewee's collaborative experiences were very limited. He asserted that the culture of his company does not look favourably on collaboration, especially collaboration with competitors. However, he does see an enormous amount of room for improvement in the company's national distribution network because many trucks have empty return trips. The company's very limited collaborative experience took place in another country and was never transferred to other regions. In particular, the experience was not transferred to the location where the Director worked because it coincided with the implementation of another project based on a perfect synchronization of the flows for goods, which could be at risk in case of a non-successful horizontal collaboration experiment. The experience the company had abroad consisted of fostering cooperation among its suppliers, some of which were very large, fast-moving international consumer-goods companies. The interviewed retailer encouraged them to jointly load the trucks with consolidated goods that were to be shipped to its hub. Our interviewee also showed special interest in promoting back hauling with suppliers, which would be vertical collaboration, but nevertheless a first step toward breaking through the company's barrier with regard to sharing information and opening itself to cooperation. He believes that horizontal collaboration is a practice that will be a must someday, probably propelled by the environment—e.g., specific laws mandating reductions in the "footprint" of their supply chain—but the company would still have to overcome its unfavourable bias against external cooperation. Then, he thought, they could explore very carefully the companies with whom they might possibly start an alliance, investigate how to do it in terms of operational compatibility and type of collaboration, and clearly define the partners' responsibilities in order to establish and maintain a highly synchronised supply chain.

5.1.6 Retailer B: Supply Chain

The last interviewee was the Director of Supply Chain Development and Projects with the same retailer as the previous interviewee.

Whereas his colleague from the Procurement team was open to cooperation, this interviewee showed a mostly reluctant attitude towards collaboration. In his eyes, it would be like publishing the strategy of the firm. However, he had been involved in very few collaborative experiences, and he underlined that his reluctance was due more to cultural issues than to operational aspects. He seemed to overestimate the disadvantages of sharing supply chain practices with other big companies, while seeing few advantages of collaboration at this time in the country in which he works. Nevertheless, he did acknowledge his interest in two specific outcomes: being a "top-to-top" company, with the visibility it implies, and learning from other companies that are more efficient in some aspects of distribution. He explained that the former outcome has become more and more relevant in the past few years because being a company that innovates and is categorized as the "top of the top" in supply chain excellence is a very efficient marketing tool. Talking about the latter outcome, he described a very concrete example in which they would search for a potential partner and investigate what that partner would offer in return for the knowledge transfer. His company has a centralized cross-docking warehouse located 550 km away from the area he is interested in learning about and from which it serves the products directly to its superstores. A smaller retailer operates just in this region, distributing its products to the stores throughout the area. Therefore, they are very experienced in the freight flow of that particular zone. The interviewee expressed his desire to access the stock of knowledge the smaller company has in the "last-mile" distribution in this particular area. In return, he would be willing to share his company's global expertise and even some space in his warehouse from which the smaller retailer could run some of its logistical operations.

In terms of horizontal collaboration, the interviewee's department is at a lower stage of development, with relational learning needing to be started in order to overcome the barriers within the company culture. In any case, there is a driver for them to start exploring, and he admitted that there was value in the potential outcome of improved efficiency in its last-mile distribution.

5.2 Horizontal Collaboration Consortiums

5.2.1 Methodology

To validate the model, it was applied in two successful Horizontal Collaboration consortiums as described in this section. The main objective was to implement Horizontal Collaboration practices in real cases with small groups of specific companies which could be used as the basis for the future extrapolation of the results to other possible groups of companies with similar needs.

The Agro-food Industry of the Autonomous Region of Aragon (Spain) was chosen as the industry and area of action. Its potentially significant margin for improvement in logistics, and especially in the logistics of manufactured product distribution to points of sale, act as a powerful driver in this sector for the implementation of collaboration practices. Specifically, the analysis focused on producers but also representatives of other sectors such as wholesale distribution and several logistics operators who offered their experiences and perspectives. Figure 5.1 presents the basic structure of the Supply Chain for the Food and Beverage Industry.

In order to successfully implement the elements for driving successful horizontal collaboration dynamics as presented in the model in Chap. 4 of this book, the research group challenged the companies involved to deploy the different steps on a *continuous relational learning context*, first *exploring together* the potential improvements of their supply chains, then individually *assimilating* the changes

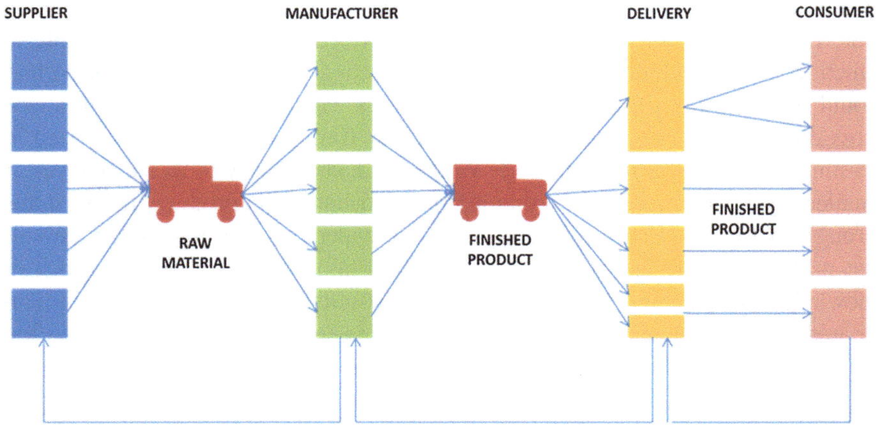

Fig. 5.1 Basic supply chain scheme for the Food and Beverage Industry

they had to make internally to enable the horizontal collaboration operations, and finally efficiently *exploiting* together the *outcomes* of their venture fairly *sharing the benefits generated*. The deployment of these learning processes allows the companies involved to create *Absorptive Capacity* in order to benefit from the potential within the horizontal collaboration relationship. This relational context leverages this potential because it implies cooperative interactions with partners during exploration as well as during the exploitation of novel boundary spanning processes, operations or products.

5.2.1.1 Data Gathering

In order to conduct the study, companies were sought with similar profiles as far as their logistics needs. In this sector, producers of perishable products and non-perishable products were differentiated given that the different requirements involved with transporting these products would make horizontal collaboration difficult.

Due to their local importance as extracted from the economic data consulted within this industry, the Meat Industry (as a producer of perishable goods) and the Wine Industry (non-perishable goods) were pre-selected as sub-sectors in order to focus the diagnostics. However, the participation of other companies from other sub-sectors that, due to their characteristics, could share transport resources with the producers of the pre-selected sub-sectors was accepted and also considered interesting.

The work was divided into three blocks in this phase:

- Three companies were contacted for the first phase and then interviewed: one from the Meat Industry sub-sector, another from the Miscellaneous Food sub-sector and the third from the Wine sub-sector. These companies had to be large as far as their logistics volume and be clearly willing to innovate.
- In a second block, companies were contacted from sub-sectors that use a transport type that is compatible with the first ones, such as Fish, Dairy, Oil or Water, among others.
- After several open days for the presentation of results and joint reflection by various stakeholders involved in the Agro-food supply chain, a third block of consultation was opened up.

5.2.1.2 Data Gathering 1: Semi-structured Interviews

In an effort to better understand the sector and the possibilities offered, three companies were interviewed from different sub-sectors. The data and information gathered from these interviews provided for the first step in approaching the exploration process, as per the model fostered by this research, as it allowed for the detection of the organizational and operational conditions of the potential consortium partners. Their answers and opinions in response to the questions posed in a semi-structured interview made it possible to focus the rest of the study.

Wine sub-sector company
A medium-size company whose principal activity is wine bottling and marketing. Its main clients are wholesale distributors and specialized companies located throughout the world. The demand they faced was variable with more and more frequent, yet smaller orders which made it very difficult to take advantage of any synergies within their own transportation flows. Their production was continuous and although most of their turnover, around 80 %, came from exports to the whole world, they had not detected any possibilities for collaboration in this area given that, in this case, they were working with full truck loads. However, they did feel that collaboration for national distribution could be useful given that they were never able to fill the trucks, due to the pattern of orders from their main retailers. According to their Logistics and Procurement Manager, this type of collaboration would be "easier and faster" with their direct competitors which were also geographically close to them but they believed this collaboration would be very fragile. Therefore, it would be more feasible to do so with non-competitor companies which offered products that were compatible with theirs from a logistics perspective (such as no need for cold-chain, no odors…). Their main drivers for the collaboration were reduced costs, a lower environmental impact and innovation. The main barriers were their own organizational culture and a lack of trust in the exchange of commercial data.

Meat industry sub-sector company
In this case, we interviewed the Managing Director of a large company dedicated to the production, preparation, packaging and distribution of meat products. Most of their turnover came from fresh products with such specific transport requirements and certain legal restrictions that, in their opinion, would make it impossible to share transportation resources with other non-meat products. With less turnover yet still outstanding and with an "upward trend", they also distributed packaged products subject to somewhat less restrictions as far as the special transport characteristics but still with some legal restrictions. Their production was continuous and their distributions, 50 % national and 50 % export, were done using the services of several transportation firms. They believed collaboration could be interesting although they also did not think it was feasible to collaborate with the competition given the characteristics of this sector.

Miscellaneous food sub-sector company
Most of their turnover came from national distribution, hypermarkets and wholesale distributors with a small yet outstanding percentage of their capacity dedicated to exports. This medium-sized company mostly worked in national distribution and exports with already full truck loads yet there continued to be a percentage of less than full truck load distributions. For the latter, they would use the services of a logistics service provider and this is where they detected a possible area of collaboration as long as the main objective was to reduce transport costs but also to improve the service. In the opinion of their Logistics and Sales Manager, the main barrier they would encounter when collaborating would be the difficulty of adapting to the delivery dates and time-windows imposed by their customers.

The main conclusion drawn from the interviews conducted was that the most important barrier to overcome when engaging in these practices would be a lack of trust and apprehension at exchanging information with competitor firms. Another factor to be considered due to the characteristics of the sector were the special transport requirements these types of products needed and how to combine them.

The most highly valued driver in all of the cases was lowering costs although others were discussed such as an improved service level or lowering CO_2 emissions.

5.2.1.3 Data Gathering 2: Surveying

In accordance with the data and conclusions obtained from the interviews, a survey was prepared asking the firms for specific data relating to their transport networks. The objective was to gather as much data as possible on their complete supply chain from procurement to distribution. Data were requested on suppliers, customers, days and points of reception and delivery, their frequency, costs, etc. The survey, which can be consulted in Annex II, was distributed to a large number of companies in the Agro-food Industry of Aragon and the data obtained were used to identify possible national collaboration routes as will be explained.

5.2.1.4 Data Gathering 3: Specific Consultations

Following a data processing phase and with possible real synergies already identified, the results were presented over several days in different forums after which a new phase of more specific consultations was opened up aimed at obtaining information from the companies relating to now specific study scenarios around horizontal collaboration.

Data were obtained from 21 companies through the three consultation phases. For the reasons already indicated, this companies belonged to a sector open to collaboration and were also solvent with the infrastructure as well as the organizational pre-conditioners necessary to carry out collaboration practices. This is why the sample was considered sufficiently representative for the study. The Figs. 5.2, 5.3 and 5.4 show the descriptive data for the sample.

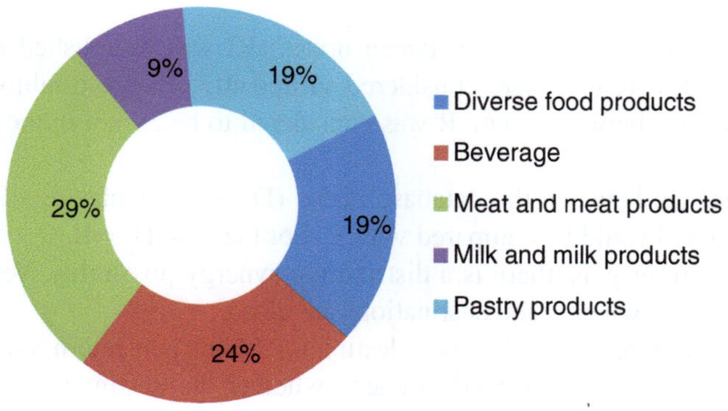

Fig. 5.2 Companies' main activities

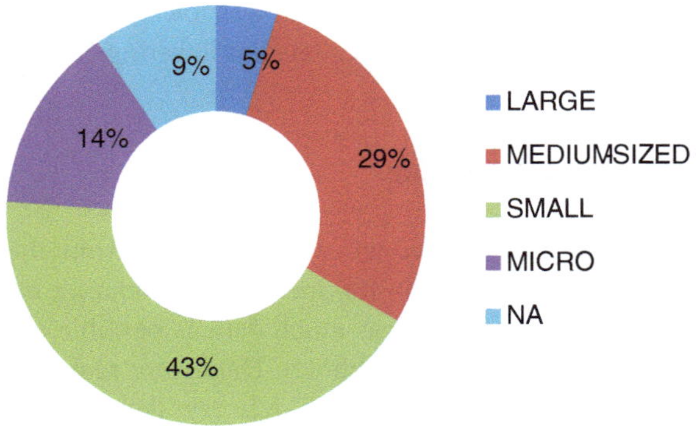

Fig. 5.3 Size of the companies according to the number of employees

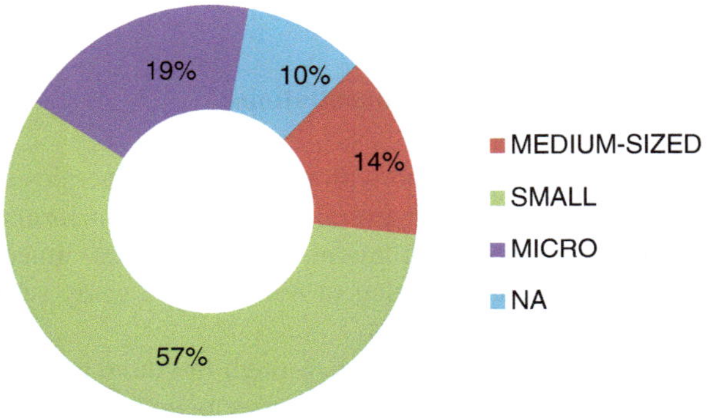

Fig. 5.4 Size of the companies according to annual invoicing

5.2.1.5 Data Analysis

As the first step in the study, criteria were defined to search for possible transportation synergies among the network nodes (example O1 and O2 in Fig. 5.5). The following procedure was followed in order to do so:

A maximum radius of distance between nodes (R) was established after which point the nodes are no longer considered sufficiently close to establish transportation synergies between them. R was considered to be 30 km in the subsequent study.

Once the distance between the database nodes (D) was calculated, all of the distances obtained could be compared with R. For the case D<=R.

If both nodes are origins, there is a distribution synergy possibility. Verification is needed to check whether the destinations are close.

If one is an origin and the other is a destination, there is a return synergy possibility yet verification is needed to check whether the destinations and origins, respectively, are close (examples O1 and D32; O3 and D13).

If both are destinations, there is a distribution synergy possibility yet verification is needed to check whether the origins are close (example D21 and D11).

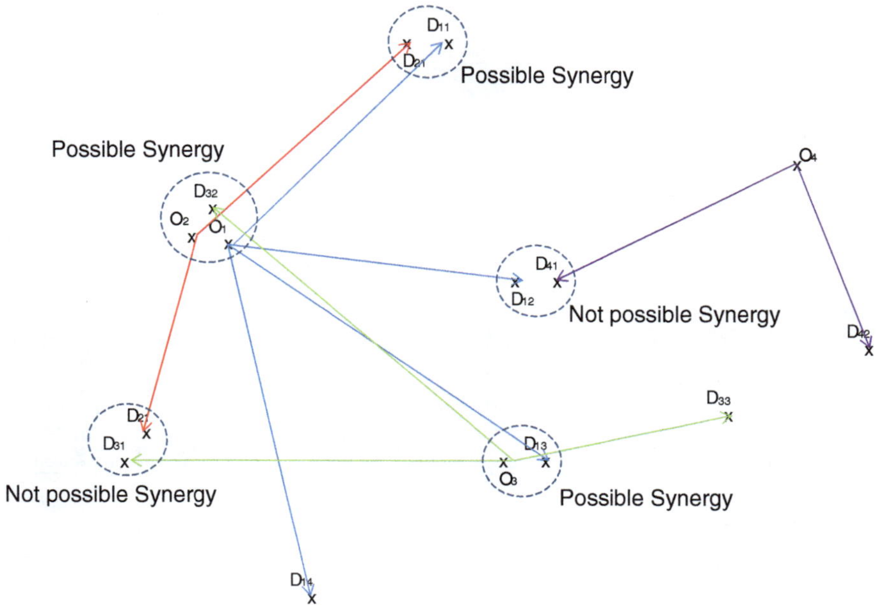

Fig. 5.5 Horizontal collaboration search criteria

Once the search criteria were established, the data analysis procedure was designed (Fig. 5.6):

Data analysis 0: Faced with the entrance of a new partner, the first step was to assign a code to said partner as well as to each one of its customers and suppliers which provides for full confidentiality as concerns the data processed. Data confidentiality was a major concern to the companies participating in this phase, partly due to the type of data handled and partly because they were being shared with competitor firms. Once the data were coded, they were reviewed and any content not relevant to the study, such as customers or suppliers without delivery addresses which made it impossible to establish shared routes, were eliminated and all of the other data were transferred to the shared database for processing and comparison with the already existing data on other partners.

Data analysis 1: Using the database, the first step was to generate each partner's own transportation networks. The data corresponding to each partner's transportation network were first imported to Google Earth and then individual supply and distribution routes were generated manually for each one (Fig. 5.7).

Once the routes were represented, the origin-destination distances for each partner were calculated. The distance matrix (Matrix 1) made it possible to quickly detect possibilities for improving each partner's route by taking advantage of their own synergies. As can be seen in Table 5.2 for example, partner 3 had three customers on one side and four on another that could be grouped and treated as

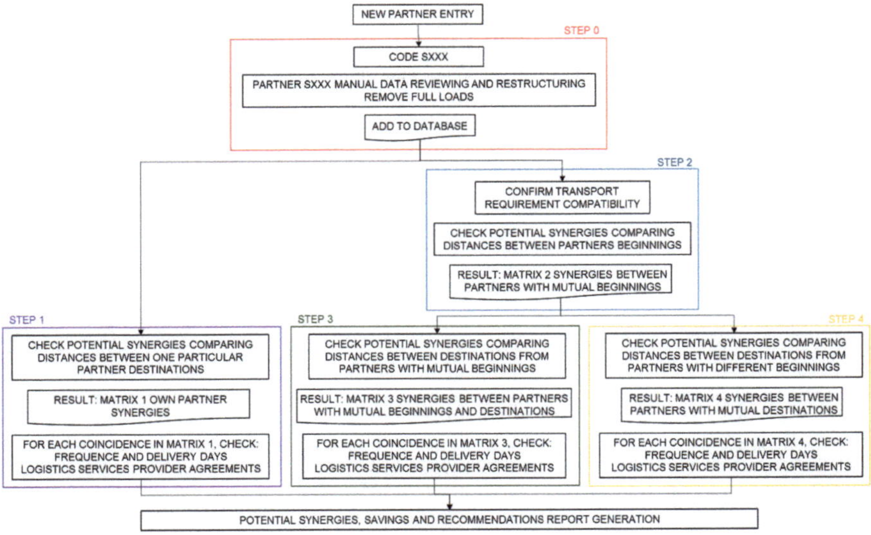

Fig. 5.6 Data analysis procedure scheme

Fig. 5.7 Own routes map (partner S003)—sample

two single nodes within their network and not as independent customers in order to optimize their transport costs.

With the combined information from the individual route map on the one hand and from the distance matrix on the other, the possible routes were obtained that each partner could attempt to exploit in order to shorten the distance travelled for

Table 5.2 Matrix 1: distance between own destinations (partner S003)—sample

	S003	S003C003	S003C001	S003C009	S003C013	S003C012	S003C006	S003C008	S003C002	S003C014	S003C016	S003C010	S003C020	S003C004	S003C018	S003C007	S003C011	S003C017	S003C005	S003C015	S003C019
S003	0	290	304	238	271	320	337	337	328	380	396	514	654	782	827	853	858	858	865	865	961
S003C003	202	0	502	526	559	388	401	401	519	661	113	795	713	571	886	912	855	855	860	860	956
S003C001	304	591	0	74,6	38,5	621	637	637	373	425	696	559	813	1.083	969	992	1.017	1.017	1.024	1.024	1.120
S003C009	238	593	75	0	43,1	555	572	572	369	388	631	522	777	1.017	932	955	981	981	988	988	1.083
S003C013	271	593	39	41,9	0	588	605	605	369	392	664	526	781	1.050	936	959	985	985	991	991	1.087
S003C012	323	387	623	557	590	0	20,6	20,6	334	401	388	435	333	608	505	531	537	537	543	543	639
S003C006	337	401	637	571	604	20,8	0	0	348	377	401	412	314	614	487	513	518	518	525	525	621
S003C008	337	401	637	571	604	20,8	0	0	348	377	401	412	314	614	487	513	518	518	525	525	621
S003C002	328	609	373	336	340	332	363	363	0	62	715	196	449	938	606	629	653	653	660	660	756
S003C014	380	661	425	388	392	380	379	379	61,2	0	767	142	465	988	552	576	669	669	676	676	772
S003C016	397	110	697	631	664	391	404	404	714	766	0	800	716	456	888	915	830	830	835	835	931
S003C010	501	795	546	521	525	414	413	413	182	141	797	0	370	1.023	413	436	563	563	570	570	631
S003C020	651	715	816	779	782	335	316	316	451	466	715	369	0	928	145	206	260	260	266	266	362
S003C004	785	625	1.085	1.018	1.051	606	613	613	937	988	458	1.023	925	0	1.098	1.124	927	927	933	933	1.028
S003C018	821	885	968	931	935	505	487	487	604	551	885	412	148	1.098	0	57,3	166	166	173	173	195
S003C007	848	912	993	956	959	532	514	514	629	575	912	437	208	1.125	58	0	215	215	222	222	240
S003C011	853	854	1.017	980	984	536	518	518	653	668	828	563	259	928	164	214	0	0	11,6	11,6	102
S003C017	853	854	1.017	980	984	536	518	518	653	668	828	563	259	928	164	214	0	0	11,6	11,6	102
S003C005	862	862	1.027	990	994	546	528	528	662	677	836	572	269	936	174	223	11,1	11,1	0	0	101
S003C015	862	862	1.027	990	994	546	528	528	662	677	836	572	269	936	174	223	11,1	11,1	0	0	101
S003C019	956	956	1.121	1.084	1.087	639	621	621	756	771	930	631	363	1.030	232	239	102	102	100	99,9	0

deliveries. This will allow them to increase efficiency and lower costs all while making more sustainable deliveries by lowering their CO_2 emissions.

For example, one of the possible routes for partner 3 could be the one outlined below which would shorten the kilometers travelled by 82 % (Table 5.3):

Besides the proximity of the transportation nodes, another important element required to gain a benefit is operational compatibility. In order to have this, the frequencies and days of delivery to each customer must be taken into consideration as they may have to be adjusted in some cases. The deliveries must also be studied to see whether they are done with a single logistics service provider or several as this would have to be unified if the route were to be consolidated. A table of recommendations was generated for each partner similar to the one below (Table 5.4):

Data analysis 2: Once the opportunities for improvement for each individual partner were detected, the study focused on detecting the synergies between the different partners which would give rise to possible horizontal collaboration. The companies subject to collaboration must be operationally compatible and so the data were divided into three groups as per the transport requirements of the products and the origin-origin distances between partners were calculated (Tables 5.5 and 5.6).

The frozen product transportation group was formed by only 1 company meaning there was no horizontal collaboration possible for this company in this study, according with the data facilitated.

Data analysis 3: Once the partners with similar transportation requirements and possibilities for collaboration due to the proximity of their origins were identified, the destination-destination distances between partners would be calculated two

Table 5.3 Own routes data (partner S003-route 004.1)—sample

ROUTE S003R004.1				
Nodes	Customers	Initial route (customers one by one)	Grouping customers in nodes (d < 30 km)	Grouping customers in routes
S003N001	S003C012	320.00	320.00	320.00
	S003C006	337.00	20.60	20.60
	S003C008	337.00	0.00	0.00
S003N002	S003C011	858.00	858.00	518.00
	S003C017	858.00	0.00	0.00
	S003C005	865.00	11.60	11.60
	S003C015	865.00	0.00	0.00
	S003C019	961.00	961.00	101.00
Total distance (km)		**5,401.00**	**2,171.20**	**971.20**
Savings in km			−3,229.80	−4,429.80
Savings in (%)			−59.80 %	−82.02 %

Table 5.4 Own routes recommendations (partner S003)

Route	Frequency	Delivery days	Own fleet versus logistics provider	Diversion from main route needed	Modification of main route needed
S003R001	Compatible	Compatible	Not compatible (3 different logistics providers)	No	No
S003R002	Compatible	Compatible	Not compatible (3 different logistics providers)	No	No
S003R003	Compatible	Compatible	Not compatible (2 different logistics providers)	No	Yes
S003R004.1	Compatible	Compatible	Not compatible (3 different logistics providers)	No	No
S003R004.2	Compatible	Compatible	Not compatible (2 different logistics providers)	No	No

Table 5.5 Matrix 2.1: distance between partner origins—dry transportation

	S002	S006	S008	S009	S013	S014
S002	X	9.5	62	62	112	101
S006		X	61.3	61.3	109	101
S008			X	0	169	40.1
S009				X	169	40.1
S013					X	209
S014						X

Table 5.6 Matrix 2.2: distance between partner origins—cold transportation

	S001	S003	S004	S006	S010
S001	X	156	160	161	156
S003		X	11.2	10.2	4.6
S004			X	21.9	6.6
S006				X	17.7
S010					X

by two. A distance matrix (Matrix 3) was obtained for each pair to again quickly detect the possible shared routes (Table 5.7).

With the combined information from the route maps on the one hand and distance Matrix 3 on the other hand, the possible shared routes were again generated in this step for partners two by two. Just to give an example, one of the routes corresponding to partners S008 and S009 is shown below (Table 5.8).

Table 5.7 Matrix 3: distance between destinations (partners S008 and S009)—sample

	S009	S009C005	S009C010	S009C012	S009C014	S009C015	S009C020	S009C023	S009C016	S009C008	S009C001	S009C009	S009C006	S009C021	S009C011	S009C017	S009C013	S009C007	S009C022	S009C003	S009C018	S009C019	S009C004	S009C002
S008	0	53	53	53	53	53	53	53	227	243	263	267	286	286	318	329	286	325	352	359	361	372	637	818
S008C022	54	5	5	5	5	5	5	5	182	274	316	320	317	317	349	360	241	274	300	308	392	425	592	849
S008C015	218	172	172	172	172	172	172	172	9	356	480	484	399	399	431	442	140	445	472	479	474	589	448	924
S008C002	263	315	315	315	315	315	315	315	488	377	0	20	355	355	367	361	547	363	343	350	381	100	804	657
S008C026	285	337	337	337	337	337	337	337	511	383	15	34	360	360	372	366	570	385	364	371	386	86	810	659
S008C021	283	235	235	235	235	235	235	235	410	502	252	255	544	544	577	588	469	116	95	102	623	360	819	899
S008C028	284	315	315	315	315	315	315	315	408	48	354	359	3	3	36	40	322	588	615	621	72	439	451	530
S008C017	313	344	344	344	344	344	344	344	437	71	377	382	29	29	18	22	350	618	644	652	44	462	469	512
S008C009	308	338	338	338	338	338	338	338	431	65	367	372	30	30	8	12	345	612	640	647	59	453	474	502
S008C011	308	338	338	338	338	338	338	338	431	65	367	372	30	30	8	12	345	612	640	647	59	453	474	502
S008C001	314	345	345	345	345	345	345	345	438	72	368	373	37	37	0	11	352	619	645	653	61	454	481	504
S008C003	303	334	334	334	334	334	334	334	426	61	373	379	26	26	14	18	341	607	634	641	61	459	469	509
S008C007	335	288	288	288	288	288	288	288	141	394	597	601	381	381	414	418	67	562	588	595	454	706	305	847
S008C027	349	301	301	301	301	301	301	301	476	568	339	342	611	611	643	654	535	54	9	17	686	447	886	986
S008C016	353	305	305	305	305	305	305	305	480	572	344	348	615	615	647	658	539	58	5	8	690	453	889	991
S008C006	360	312	312	312	312	312	312	312	487	579	351	355	622	622	654	665	546	63	9	0	697	460	896	999
S008C025	352	404	404	404	404	404	404	404	577	418	80	102	408	408	375	369	636	451	431	438	390	60	848	666
S008C010	365	317	317	317	317	317	317	317	491	584	370	374	626	626	659	670	551	59	38	33	702	479	901	1018
S008C019	368	320	320	320	320	320	320	320	494	587	373	377	629	629	662	672	553	62	41	35	704	482	904	1020
S008C013	447	400	400	400	400	400	400	400	253	461	708	713	451	451	486	491	178	673	700	707	525	818	188	824
S008C024	507	562	559	559	559	559	559	562	735	429	238	257	417	419	384	380	717	606	588	593	398	174	856	518
S008C023	518	472	472	472	472	472	472	472	383	377	688	693	329	329	364	370	255	745	772	779	404	773	122	659
S008C012	695	725	725	725	725	725	725	725	818	452	504	526	417	417	388	382	732	875	855	862	364	442	861	242
S008C008	809	840	840	840	840	840	840	840	913	567	656	661	532	532	503	497	786	1010	990	997	479	681	775	0
S008C020	817	848	848	848	848	848	848	848	917	575	664	669	535	535	511	505	790	1018	998	1005	487	687	779	4
S008C014	822	853	853	853	853	853	853	853	925	580	669	674	542	542	515	509	797	1023	1003	1010	491	676	787	14
S008C004	810	841	841	841	841	841	841	841	934	568	620	642	533	533	504	498	848	991	971	978	480	558	976	205
S008C018	816	847	847	847	847	847	847	847	940	574	626	648	539	539	510	504	854	997	977	984	486	564	982	211
S008C005	835	789	789	789	789	789	789	789	700	646	957	962	599	599	633	639	573	1062	1089	1096	673	1043	267	928

Taking advantage of these synergies depends on the willingness of the partners to collaborate and the possibility of modifying or adjusting some of the parameters of their distribution networks, such as the frequencies and days of delivery to each customer or the logistics service provider used (Table 5.9).

Possibilities for collaboration were found for S002-S006, S008-S009 and S003-S004 by crossing the coinciding partner origins and destinations.

Table 5.8 Collaborative route data (partners S008 and S009 route 002.1)—sample

ROUTE S008S009R002.1				
Nodes	Customers	Initial route (customers one by one)	Grouping customers in nodes (d < 30 km)	Grouping customers in routes
S009N003	S009C008	243.00	243.00	243.00
	S009C006	286.00	286.00	48.40
	S009C021	286.00	0.00	0.00
S008N001	S008C028	285.00	285.00	3.10
	S008C009	311.00	28.50	28.50
	S008C011	311.00	7.90	7.90
	S008C001	318.00	15.70	15.70
	S008C017	319.00	14.70	14.70
S009N004	S009C011	318.00	318.00	0.00
	S009C017	328.00	10.70	10.70
	S009C018	361.00	361.00	52.70
	S008C012	703.00	703.00	364.00
S008N004	S008C026	822.00	822.00	125.00
	S008C018	825.00	617.00	617.00
Total distance (kms)		**5,716.00**	**3,712.50**	**1,530.70**
Savings in km			−2,003.50	−4,185.30
Savings in (%)			−35.05 %	−73.22 %

Table 5.9 Collaborative route recommendations (partners S008 and S009)—sample

Route	Frequency	Delivery days	Own fleet versus logistics provider	Diversion from main route needed	Modification of main route needed
S008S009R001.1	Not compatible	Not compatible	Not compatible (3 different logistics providers)	No	No
S008S009R001.2	Not compatible	Not compatible	Not compatible (3 different logistics providers)	No	No
S008S009R002.1	Not compatible	Not compatible	Not compatible (3 different logistics providers)	No	No
S008S009R002.2	Not compatible	Not compatible	Not compatible (3 different logistics providers)	No	No

Data analysis 4: Possible collaboration between companies with different origins; in other words, origins at a distance of more than 30 km, was also analyzed so as not to miss any opportunities. Some additional possibilities subject to exploration were obtained by conducting a similar study to the one outlined previously: S002-S008, S002-S009 and S002-S014.

5.2.2 Successful Case: Food Industry

In view of the results once the data for all of the partners were compared, the research group focused on the first pilot study around the possible combinations that provided for transport collaboration for the dry products coming from pair of partners S002-S006-S008-S014. The activities of the four firms chosen ranged from the production and commercialization of baked goods and pastries to the production and commercialization of pasta and the bottling and commercialization of wine. They were companies of a diverse size, all with continuous production where the smallest worked with their own fleet and the largest used their own trucks as well as the services of a carrier or logistics service provider. As far as their markets, one of them mainly distributed abroad (80 %) where the other three dedicated said percentage of their production to the national market.

All of them had a priori expressed willingness to collaborate with companies in their same sector mainly because of the special characteristics of their food products. It was at this point when the research team explored the possible compatibilities between the partners and the inter-organizational learning processes began between them, namely exploration, assimilation and exploitation, in order to create Absorptive Capacity from the existing relationships within the consortium, as described in Chaps. 2 and 4 of this book. Let's describe below how each of these three learning processes was deployed within the development of the horizontal collaboration practices.

5.2.2.1 Exploration

This first learning process, exploration, refers to recognizing and understanding potentially valuable new knowledge coming from the horizontal collaborative partner through exploratory learning. The data analysis previously performed by the research group was used to ensure that this process would be carried out efficiently. Each partner's individual routes were superimposed with those of the others and finally 8 routes with collaboration possibilities were found nationally in Spain:

Route 1. Barcelona-Gerona Zone
Route 2. Levante Zone (Valencia)
Route 3. Levante Zone (Murcia)
Route 4. Central Zone (Madrid)

Route 5. Southern Zone (Malaga)
Route 6. Southern Zone (Seville)
Route 7. West-Central Zone
Route 8. Northern Zone

Although not all of the partners covered all of the routes, the coalition formed by the consortium of the 4 partners did cover all of them meaning it was possible to generate synergies in any of them. The optimal combined routes are represented graphically in the maps of Annex III following the cost reduction criteria. As an example, the one corresponding to the Central zone (Madrid) is shown below. Three of the consortium partners coincide on this route (S006, S008 and S014) with different origins and a common destination although with diverse delivery points (Fig. 5.8).

Once the possible collaboration routes were identified and each partner's customers were included on that routes, the frequencies and days of delivery of each partner to each one of their customers were reviewed in an aim to evaluate their operational compatibility. Due to the variability of the data in relation to these two parameters, a theoretical annual delivery calendar was established for each route. To calculate the theoretical costs of each one of the routes, isolated orders or those due to special promotions which showed no frequency in time were eliminated from the calendars (Table 5.10).

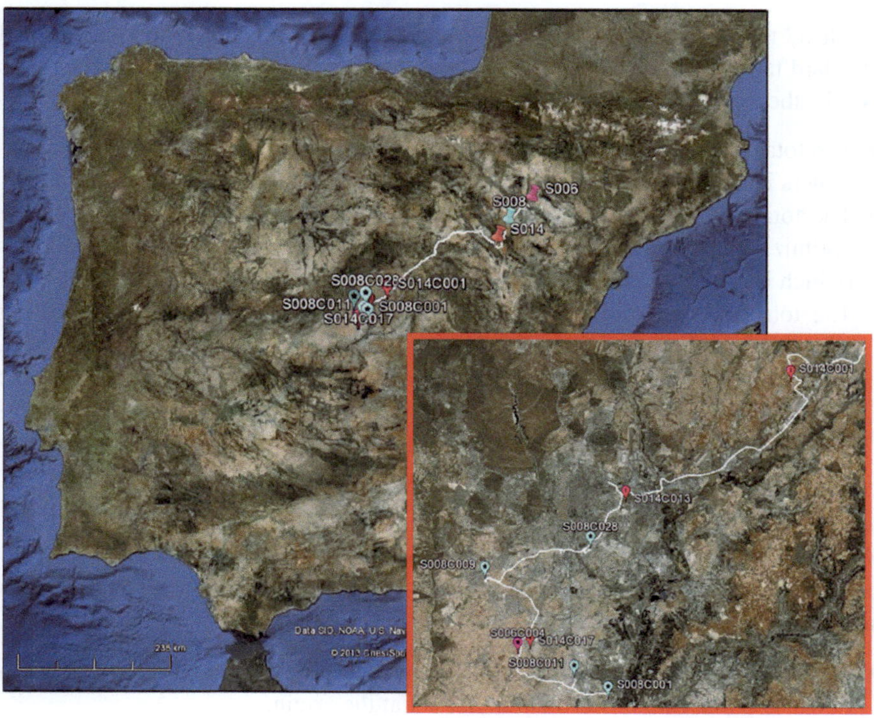

Fig. 5.8 Collaborative route map (Madrid area)—sample

Table 5.10 Theoretical annual deliveries (Madrid route)—sample

Customer (node)	Delivery frequency	Delivery day	Per month							
			Per week		Per week		Per week		Per week	
			d1	d4	d1	d4	d1	d4	d1	d4
S014C001	Once per month		3							
S014C013	Once every two months		8							
S008C028	Once every two months		1							
S008C009	Twice a week	Wednesday and friday	12	12	12	12	12	12	12	12
S014C017	Once per month		1							
S006C004	Once every two weeks		3				3			
S008C001	Once a week	Tuesday	1		1		1		1	
S008C011	Once a week	Friday	9		9		9		9	
Total per day (pallets)			38	12	22	12	25	12	22	12
Total per month (pallets)			50		34		37		34	
Total annual (pallets)			**1,800**							

Next, the following parameters were calculated based on a pre-established standard tariff for each one of the routes and for each one of the delivery dates as per the theoretical calendars determined:

- The total transportation cost if each partner were to serve each one of their customers individually (which is referred to as "node to node").
- The total transportation cost if each partner were to serve their customers by optimizing their own routes; in other words, by grouping their own customers (which is referred to as "own routes").
- The total transportation cost if the partners were to work together by taking advantage of the synergies offered by the collaborative network. In this case, the idea is always to optimize the transport costs meaning the lowest possible cost is always considered in the calculations which is either the cost of the combined route (including customers of different partners in a single transportation resource) or the cost of the individual routes (with each partner working individually).

When doing the calculations and using a theoretical standard tariff to do so, various hypotheses were formulated resulting in several possible cost and savings scenarios:

- Transport costs using a full truck tariff and assuming a fixed sum for each additional stop made at the destination but not at the origin.
- Transport costs using a full truck tariff and assuming a fixed sum for each additional stop made both at the destination and at the origin.

- Transport costs using a partially loaded truck tariff and assuming a fixed sum for each additional stop made at the destination but not at the origin.
- Transport costs using a partially loaded truck tariff and assuming a fixed sum for each additional stop made both at the destination and at the origin.

The results of the potential theoretical economic savings that could be obtained with the collaborative network are shown in Table 5.11.

By using the same transport to distribute their products either because they share customers or because they share destinations, the total annual savings the 4 partners could obtain in their transport costs by working together would be between 6.5 and 17 %. These figures have been proved a posteriori by carrying out the real operations.

As can be seen, considering additional stops or not at the origin given that the route origins are different generates an additional cost that must be taken into consideration when calculating the costs attributable to each route. The influence of this cost on the potential savings was checked in the study by analyzing their variability with different unitary price estimates per additional stop within a commercial range (Table 5.12):

Table 5.11 Potential theoretical economic savings within the collaborative network according to different scenarios

Scenarios	Annual pallets delivery	% of savings in comparison with own routes
a. Full truck tariff without any additional cost due to stops in origin	4.053	−16.88
b. Full truck tariff with additional costs due to stops in origin		−13.06
c. Less than a truck tariff without any additional cost due to stops in origin		−11.49
d. Less than a truck tariff with additional costs due to stops in origin		−6.51

Table 5.12 Potential theoretical economic savings within the collaborative network taking into account additional costs due to each additional stop in the main route

Scenarios	% of savings in comparison with own routes			
Additional cost due to each additional stop in the main route	30 €	35 €	40 €	45 €
a. Full truck tariff without any additional cost due to stops in origin	−16.88	−17.41	−17.94	−18.47
b. Full truck tariff with additional costs due to stops in origin	−13.06	−12.96	−12.86	−12.76
c. Less than a truck tariff without any additional cost due to stops in origin	−11.49	−10.26	−9.06	−7.87
d. Less than a truck tariff with additional costs due to stops in origin	−6.51	−4.77	−3.48	−2.47

The results obtained, potential savings for all of the routes and possible collaboration scenarios were presented to the selected firms individually in order to establish their individual willingness to collaborate after internally analyzing their own processes and the advantages this collaboration could offer them in terms of lower costs and improved efficiency; in other words, their potential relational rents, in accordance with the horizontal collaboration model presented in Chap. 4.

Once accepted by all of the parties and their willingness to collaborate was individually expressed, a joint working session was conducted. The consortium firms held several meetings together which made clear the trust each had in the others, as they were all solvent companies within the sector and some of them of renowned prestige, as well as the organizational compatibility among them, as in all cases the conditions of a corporate culture and business philosophy geared towards innovation and the implementation of actions such as those presented in this research work were fulfilled. Moreover, they visited a plan for wine bottling in an effort to develop a relaxed and trustworthy working environment. The partners themselves decided upon the terms of their collaboration at this and later sessions.

The decision was made to start with short-term cooperation (Lambert et al. 1999) and choose a single destination from all of the possible ones to do a first joint pilot shipment, in which all the initial agreements were deployed. The partners chose the Madrid route scenario to make this first real joint shipment. As per the element of mutuality, the transport resource to be used would be partner S002's own fleet. This implied that they had to get confirmation that partner S002's insurance policy would cover the full load.

5.2.2.2 Assimilation

The second learning process, assimilation, involves the processing of valuable new knowledge extracted from the exploration process together with the other horizontal collaborative partner, and integrating it with each of the individual firm's existing knowledge. To this end each partner in the consortium had to internally review and align their processes seeking the operational symmetry that would enable horizontal collaboration to design the specific operations for the shipment. The steps to follow were defined one by one and the decision was made for all communications to be via email with a copy to all of the parties involved. The information to be exchanged weekly would be:

- Each partner's planned deliveries to Madrid
- Collection and delivery time windows each had agreed to with their internal customers
- Transportation costs (which changed each week depending on the size of the load and what each partner contributed)
- The final shipment confirmation.

It was also made clear that the element of solidarity had to exist among the partners in order for the collaboration to work as partner S006 had a more restrictive time window than the others for one of their deliveries and the joint operations were adapted so this delivery could be completed without any problems. The implication of this aspect on our model was that all of the other partners were aware of the limitations of this particular delivery and the corresponding agreement was reached within the consortium so the problem could be resolved.

One again, a way to get the imperative element of trust, that enables collaboration to develop, was for each one to know very clearly what their particular role was, as well as those of their partners within the coalition from the very beginning. They defined, discussed and agreed to these four roles:

- The shipper, which would correspond to each one of the manufacturers/partners collaborating
- The on-line neutral intermediary (or trustee), which was responsible for the operations and, in our case, this role was filled by one of the shippers
- The off-line neutral intermediary (or trustee) which actually performs the work of the neutral external element, a responsibility that fell on the research team doing this study
- The logistics service provider, which was not present in our pilot shipment since the transportation service was provided by the own fleet of one of the shippers, but would need to be considered for future pilot studies.

The type of contract chosen to gather the agreements reached in the end was a document drawn up throughout the different sessions, which was distributed a posteriori in order to be expressly accepted and serve as a letter of intent before the actual activities began. A commitment by all of the parties to comply with the agreements reached and certain rules agreed to beforehand for possible breaches are essential to the sustainability of the collaboration.

One of the most important agreements reached was how to establish Value Sharing, in line with the principles presented in Chap. 4. By being responsible for the transport with their own resources, partner S002 had to inform the others prior to the shipment of the tariff that would be applied to each one. This had to be based on the each partner's individual transport costs and the distribution of the savings established as agreed. It was necessary for all of them to establish and agree to a percentage-based distribution of the savings generated for each one of the partners. Gain sharing distribution techniques were used to calculate this distribution based on game theory and, specifically, the Shapley Value (Shapley 1953). These interactions between the partners make clear the interactions of the different inter-organizational learning processes (exploration) and intra-organizational learning processes (assimilation) in their efforts to agree to the joint terms of the operation and their involvement in the gain and value sharing.

In our case, the application consisted of each possible coalition being assigned the function that corresponded to the total transportation cost and calculated based on the destination (km travelled) and the load carried (number of pallets). In the case of the coalition where the players/partners worked individually, their own tariffs were applied. However, the common tariff agreed by all was applied to the joint coalitions of 2, 3 and 4 partners.

After this, all of the distribution possibilities were generated considering each player/partner's order of entry in the coalition. For example, assuming an entry sequence of 1-2-3-4, when partner 1 enters, the transportation cost is the one it generates by itself and so there are no savings to be distributed. As soon as partner 2 adds its load to the shipment, the joint transportation cost is lower than if doing so individually and these savings are applied to player 2 and, thus, its cost or contribution to the cost is lower. When 3 enters and finally 4, the same method is followed, applying the savings generated by the coalition to the last one entering. With this process, each one of the players/partners are assigned their own marginal contribution (MC) to the group which had been formed from the time of their entry. By repeating this sequence for each one of the possible coalitions (based on the order of entry of each one) and calculating the average of each one's marginal contributions, the weight of each player/partner in the collaboration was obtained and thus the distribution percent of the savings generated that would "fairly" correspond.

Shapley's value is the only value that satisfies the properties of efficiency, monotonicity (higher value to coalition, higher payoff), dummy (no value no payoff) and individual fairness.

The application of this methodology enables a fair distribution of the savings considering each partner's contribution to the collaboration in terms of individual initial costs, the loads contributed and the distances to be travelled. Notwithstanding, the distribution proposed also took into account other factors that affect collaboration such as contributing one's own fleet or the services of their own logistics operator, modifying frequencies and/or days of delivery to adjust to the established route and contributing a stable load as far as the frequency and size of the order to the coalition. These aspects were provided for by adding certain bonuses to the partners in the distribution of the savings. The bonus percentages to be applied to the pilot shipment were agreed to by the partners at the preliminary meeting and were as follows:

- for the use of one's own fleet: 5 %,
- for the use of a logistics operator: 2 %,
- for modifying delivery frequencies, delivery days or shipment stabilities: 0 %.

5.2.2.3 Exploitation

The final learning process, exploitation, refers to using the assimilated knowledge to create new knowledge and then generating real commercial outputs in terms of relational rents, as presented in Chap. 4.

In the framework of this case and once all of the parameters of the collaboration were agreed after having studied all of the potential benefits and after each partner had analyzed and adapted their particular internal processes so the horizontal collaboration operation could take place, the actual joint shipments began through the corresponding pilot shipment. Four shipments had been made as of the date this book was written.

By looking at the relational rents obtained by the partners, the improved efficiency can be quantified as the economic savings achieved with each shipment, which falls within the theoretical bracket calculated a priori (Table 5.13):

Table 5.13 Actual savings in transport costs obtained by the partners with collaborative practices

Delivery date	Partners				Savings (%)
	S002	S006	S008	S014	
21/05/2013	X		X		−9.36
28/05/2013	X	X			−12.96
25/06/2013	X	X			−10.21
09/07/2013	X	X			−10.21

The shipments, which had been previously done in different means of transportation, were unified in a single truck which reduced each one's carbon footprint which also meant an improvement in terms of sustainability. Likewise, the companies' customer service and flexibility in complying with the ever more demanding requirements from their customers requesting more frequent deliveries of smaller quantities also improved so some of the companies could now make weekly shipments of just a few pallets.

The companies participating in this pilot test continued innovating and collaborating at the end of the study following the methodology described in Chap. 4 for the establishment of communications and the generation of agreements and using the calculation tool based on Shapley's value provided to them to calculate the tariffs and distribution of the savings. Thus, the dynamics of continuous relational learning were taking place through additional explorative, assimilative and exploitative learning processes. The model dynamics are subject to a continuous improvement loop given the dynamics of the companies' businesses which allow for relational learning to take place in that the outputs from the exploitation are the starting points for new exploration opportunities.

As of the date the results of this study were drafted, the four partners were looking for collaboration scenarios in addition to the Madrid route and were beginning to explore Valencia and Barcelona routes.

5.2.3 Successful Case: The Beverages Industry

After several presentations of the results obtained during the data analysis phase in different forums, such as the FoodForLife platform in Madrid and at the Aragon Agro-food Industry Association Assembly, among others, and with a view to exploring the advantages offered by combining efforts and loads with shared destinations, a shorter second consultation was launched for companies in the food industry to find out about their activities in 3 specific scenarios of activity. The consultation also aimed to establish the willingness of this second group of companies to participate in the preliminary study into the possibilities of collaboration and the cost savings that could be generated as well in a second pilot test.

The same search criteria and the same procedure previously described were used to search for possible collaboration synergies in this second case. The companies were solvent, renowned and had the necessary infrastructure to implement the

Table 5.14 Initial data for Madrid scenario

Partner	Load (pallets)	Delivery day	Frequency	Customer
S008	15	Wednesday and friday	Twice a week	Retailers 1 and 2
S020	10–16	Wednesday and friday	Twice a week	Retailer 1
S002	2	Tuesday and thursday	Twice a week	
S017	2		Once every two weeks	Retailer 2
S018	1	Tuesday and thursday	Twice a week	Retailer 1
S014	Full truck load			
S021	–	Friday	Once per week	Retailer 1

collaborative practices in addition to being sensitive to innovation for continuous improvement.

The analysis focused on one of the scenarios proposed in this second consultation which corresponded to a shipment to two specific big retail customers, known herein as Retailer 1 and Retailer 2, located in the same city.

The data were gathered in Table 5.14.

5.2.3.1 Exploration

Considering these data and the possibilities they offered, the decision was made to explore the joint shipment between partners S008 (a company dedicated to the production, bottling and commercialization of wine) and S020 (a company dedicated to the production, bottling and commercialization of beverages). Both of the companies were medium-sized with a significant percentage of their production dedicated to exportation. They used several carriers for their national distribution.

In this case, a single route was analyzed—the one corresponding to the Central Zone (Madrid). Taking advantage of the synergies found, the potential savings for each partner could reach up to 50 % of the total transport cost for this route. Two theoretical studies had to be designed before the parties could reach an agreement on doing the actual pilot shipment.

FIRST STUDY: the total costs and savings involved with the horizontal collaboration were calculated based on the unitary costs contributed at first by each partner and the load to be carried by each one using the services of the carrier previously used by partner S020. Considering that two un-loadings were done at the destination, the savings were calculated to be around 7.5–8 %. With this scenario, one of the companies was reluctant to initiate the collaboration so the partners were asked for more information in order to conduct a more detailed study of the possibilities which a priori seemed to offer the parties significant benefits. The barriers associated with the lack of trust and the companies' reluctance to offer data was a major obstacle in this case and almost broke the coalition.

SECOND PILOT TEST: a new scenario was designed with the additional data put forth by the partners and the calculations were re-done for the total costs and savings for each one. With these new conditions, the potential savings with each shipment were found to be around 42–49 % of the total transport cost using the full truck tariff with additional stops at the origin (since the starting point of both routes did not coincide) yet not at the destination (as it was the same customer and same location) for the calculation.

5.2.3.2 Assimilation

Again, and once the parties had individually expressed their willingness and agreed to collaborate, the agreements under which the shipments would be completed were established jointly, this time via telephone and email.

Both companies were operationally symmetrical for this route, made their deliveries to the same destination at an identical frequency and with coinciding delivery days. The only differentiating aspects were the points of origin of the route, which were different, and the carrier making the delivery which was different for each partner. Each one of the partners reviewed their processes using assimilation processes as they had to align these aspects following the elements of mutuality and solidarity among the partners so the collaboration could take place.

The decision was also made in this case to begin with short-term cooperation using the same type of contract as in the previous case, a letter of intent that was written with the express acceptance of the agreements reached via email. This is possible in these first steps of collaboration and makes clear the trust that exists between the partners; however, as the collaboration goes forward and other types of closer collaboration are initiated, the signing of legal agreements and contracts takes on even greater importance.

The services of a third party, a transport company, were used for the transportation. Before the collaboration began, both companied had been making their shipments with two different carriers. The parties decided that the joint shipment would be made by the carriers that had done them previously, rotating month from month and that each month, the transport costs would be invoiced to one of the partners. The imperative element of mutuality in this case was essential for the collaboration. A specific timeline was also established for pick-ups and deliveries.

The four horizontal collaboration roles established in this case were:

- The shipper, which would correspond to each one of the two manufacturers/partners collaborating;
- The on-line neutral intermediary (or trustee), which was responsible for the operations and, in this case, was also filled by one of the loaders;
- The off-line neutral intermediary (or trustee), which also fell on the research team doing this study;
- The logistics service provider which, in this case, were the two carriers that did the shipments each month.

To establish the value sharing and the distribution of the savings generated in a way that was as fair as possible, the same tool designed for the previous case based on the Shapley Value was used. Also taken into consideration were the other factors that affect and support the collaboration such as the fact that none of the partners were contributing their own fleet, no frequencies or delivery days were modified to adjust to the established route, both companies were 100 % operationally compatible in this aspect and the fact that both contributed a stable load as far as the frequency and size of the order to the coalition. The bonuses for these aspects were agreed upon internally by the partners.

5.2.3.3 Exploitation

The first joint shipment of this coalition brought about total cost savings for both companies of approximately 47 % of the total transport cost for this particular shipment which is a substantial improvement in the efficiency of this process for both companies. The materialization of these savings through this means of shipment was key to the sustainability over time of the collaborative practices.

Other non-quantifiable benefits included the possibility of completing the transportation process on this route in a more environmentally sustainable manner by practically cutting the total number of kilometers travelled in half and, therefore, also substantially reducing the carbon footprint. Moreover, greater flexibility with these shipments was explored whereby the companies considered doubling the delivery frequency as a way to improve the level of service they offered their customers.

Following the continuous improvement loop that the model offers and in spite of their initial reluctance, both companies continued collaborating with periodical shipments after this research work was finished which brought them significant economic savings of around the percentages mentioned as well as various other benefits.

References

Lambert DM, Emmelhainz MA, Gardner TJ (1999) Building successful logistics partnerships. J Bus Logistics 20(1):165–181

Sáenz MJ, Revilla E, Knoppen D (2014) Absorptive capacity in buyer-supplier relationships: empirical evidence of its mediating role. J Supply Chain Manage 50(2):18–40

Shapley LS (1953) A value for n-person Games. In: Kuhn HW, Tucker AW (eds) Contributions to the theory of games, volume II. Annals of mathematical studies 28, pp 307–317. Princeton University Press, Princeton

Yin RK (2003) Case study research. Design and Methods. Sage Publications Inc, Thousand Oaks

Chapter 6
Managerial Recommendations

6.1 Drivers of Horizontal Collaboration

As stated in Chap. 4, the drivers are opportunities perceived by the companies as elements they could exploit to its advantage for promoting horizantal collaboration. They come from the environment or from inside the organization and as both are in constant change, new drivers appear throughout time and others lose its exploitation potential and become less attractive. For this reason and in order to keep it as generic as possible, the model does not limit itself to a given list of unique driving forces.

Nevertheless, within the case studies presented in Chap. 5 the senior executives shared their opinion about a set of pre-defined drivers and assessed them by giving them a mark on a Likert scale, varying from 1, which meant it does not encourage companies to consider starting horizontal collaboration, up to 5, which means it strongly encourages this kind of cooperation. Analyzing the results we can take some conclusions about the most common drivers and ones that offer more potential in the eyes of the experienced managers.

Table 6.1 summarizes their input, showing the average mark given to each driver.

The most valued opportunities have been reducing distribution costs, allowing easier response to demand fluctuation, increasing the service level and optimizing the loading rates. Lower carbon emissions and accessibility to new markets have turned out to be the less valued ones.

Although there are many discussions about it, distribution is often considered as a non-value-add activity. Companies focus efforts in finding opportunities to reduce its related costs. In the context of horizontal collaboration it has even been considered by our interviewees as the main driver for starting cooperation.

Furthermore, by listening to the voice of the customer, who expects higher and higher delivery performances, it is clear that being able to react quickly to demand fluctuations and improving the service levels have become top priorities.

M.J. Saenz et al., *Enabling Horizontal Collaboration Through Continuous Relational Learning*, SpringerBriefs in Operations Research,
DOI 10.1007/978-3-319-08093-2_6

Table 6.1 Ranking of drivers for horizontal collaboration

Motivations	Assessment
Cost reduction	4.6
Allowing easier response to demand fluctuation	4.4
Improvement of the service level	4.2
Improvement of the vehicle fill utilization	4.2
Lower carbon emissions	3.2
Access new markets	3.0

Also, when reading the low fill utilization rates of the vehicles in Europe it is no wonder that the companies seek ways to optimize the load of the trucks. In this regard, horizontal collaboration is a great opportunity for them, since it offers load combinations that cannot be found within the own supply chain.

The differences between these three most valued opportunities and the other less valued ones can be attributed to the fact that the first ones are drivers shared by nearly all sectors and are essential to gain competitive advantage towards competitors in many businesses and the other two strongly depend, in the case of the footprint, on the specific legislation and environmental concerns of the company's country. In the case of the market prospection, it depends on the growth strategy and development stage of the particular firm.

In this sense, although some elements can be designated as universal driving forces, the opportunities that a company may find to start horizontal collaboration is dependent of its environment, since what today may be a key element of success for the business, might not be any more on a mid-long term perspective and new element can appear due to changes on the economic, social or legislative context of the firms.

Despite the different assessment of the given list of drivers, the executives agreed on saying that this kind of alliances can become an excellent way to exploit whatever opportunities are found. Therefore the recommendation is to be always alert to constantly find out new opportunities to gain competitive advantage and be proactive in the search for partners in order to have the possibility to jointly explore horizontal collaboration as a path to improve the distribution practices.

6.2 Barriers to Horizontal Collaboration

The logistics managers of the companies consulted in Sect. 5.1 of this book were asked to indicate the obstacles they believed existed when initiating horizontal collaboration. Just as was the case with their motivations, they were asked to rate the impact of the barriers on a Likert scale of 1 to 5 (where 1 was very low and 5, very high) (Table 6.2).

The results obtained show that the main barriers found by companies when initiating horizontal collaboration are, by order of importance, the organizational

Table 6.2 Barrier ratings

Barriers	Rating
Organizational culture	4.2
Lack of trust	4.0
Difficulty finding collaborators	3.6
Lack of common processes	3.6
Competitors acquiring information	3.2
Difficulty agreeing to HC terms	3.2
Difficulty distributing the benefits in a balanced manner	2.4

culture, a lack of trust, the difficulty finding collaborators, a lack of common processes, competitors using the collaboration to acquire information, the difficulty of agreeing to the terms of the horizontal collaboration and the difficulty of distributing the benefits in a balanced manner.

Of the three types of collaboration defined by Lambert et al. (1999), horizontal collaborations of the first type have been developed in the consortiums presented in this book; in other words, sporadic short-term cooperation for joint distribution, the management of which was ad hoc for the specific situation. The obstacles to horizontal collaboration listed in the paragraph above were made clear when doing this research work.

A lack of support by company management, which has an essential role when it comes to fostering horizontal collaboration from the highest levels of the company to the lowest levels, in many cases has led to non-participation in initiatives like the one presented herein with excuses such as a lack of resources, a lack of time to prepare the data requested adequately or a fear of worsening the service level, for example.

A lack of experience with these practices, which generates a lack of trust in competitors, creates a certain reluctance to share relevant information in fear it will be made public or could be used by the competition.

According to the studies conducted, companies will not collaborate with companies they do not trust or which have a bad reputation nor will they collaborate with weak or incompatible companies that could affect their image. This, along with the difficulty of integrating their distribution networks, makes it difficult to find collaborators with which a firm can study possible synergies.

Due to the sector characteristics in our case with customer networks that are too fragmented and result in many nodes to be served, with high yet not very cyclic delivery frequencies and low loads per shipment, there are different requirements depending on the product (cold/frozen, dry). This makes the transport of some products incompatible at times or the fragility of some products which cannot be stacked on pallets to optimize trips, finding routes that optimize each company's individual transportation processes is a complex task that requires the person doing it to have a large quantity of data and extensive knowledge of how each company's distribution networks work.

Another barrier found when establishing horizontal collaboration practices in the pilot shipments described in Chap. 5 is the difficulty of agreeing to the terms of the very collaboration. Certain partners show little implication in the operations yet they really need to be willing to collaborate on all levels on aspects such as confirming on-time loading, transmitting information fluidly and generating the necessary administrative documentation. Other aspects such as ensuring the loads and making sure someone is responsible for loading at the time of transport and delivery are critical points that must be agreed to beforehand. Heide and John (1992) already mentioned the concept of solidarity in the scope of collaboration and this was very obvious in the experiences with the case studies analyzed. Whenever the solidarity element is present in collaboration, "problems will be solved as a team and the goals will be reached together".

Finally, one major concern for some of the companies that collaborated in the pilot shipments was the method that would be used to establish the distribution of the savings generated. Achieving a balanced distribution of the benefits and making it so the companies can perceive recognition for their additional efforts in favor of the collaboration is essential to its success and sustainability over time. The literature talks about mutuality (Lado et al. 2008) where these authors believe that "even when the efforts, costs and benefits are not always distributed equitably, they are balanced out over the long term."

The results make it clear that companies find risks and face barriers associated with the implementation of horizontal collaboration practices. Thus, the challenge lies in creating mechanisms to mitigate these risks and activating the aforementioned drivers to manage sustainable, long-term collaboration.

6.3 Best Practices and Recommendations

Using the problems detected in the cases analyzed as the starting point, a series of recommendations are proposed below which make horizontal collaboration between companies feasible and sustainable and which make it possible to properly deploy the different mechanisms and learning processes described for the model in Chap. 4 and illustrated with the different case studies:

1. The entire company, from management to those actually responsible for the operations, must be involved in the collaboration. There must be an **organizational culture that is open to innovation** within the company. According to the literature consulted, the characteristics of the key stakeholders most willing to collaborate would be those with several years of experience in logistics who are specialists in this area without any other responsibilities, those who are pragmatic and prepared to share and assume risks, who do not get blocked by the little details and are familiar to actively sharing experiences in forums and benchmarking… those who are flexible, innovative and creative and also have the full support of their superiors as management plays a fundamental role when it comes to fostering these practices within a company.

2. The **type of company** to collaborate with must be very clear either because they have compatible networks with common nodes (either origins or destinations) or because nodes are shared. This makes easier the search for potential partners as well as the effective implementation of the collaboration. Moreover, the search for a partner for collaboration aims to take advantage of other factors such as the other company's know-how to access certain markets, for example. The mutual learning factor is also one of the greatest motivations behind companies setting up initiatives of this type. The **type of collaboration** to be established must also be decided. Normally, companies begin with the first type of collaboration as described in this book to later move on to a second level where the companies are willing to share more operational information and are capable of establishing joint and synchronized medium-term plans and finally reach the last type where the relationship transforms into a long-term collaboration type where the parties' logistics departments end up becoming extensions of each other.
3. Before creating horizontal collaboration practices between companies, each company must verify both their **available resources** as well as their operational **compatibility** against those of their potential collaborators. This compatibility between the different human resources, the key stakeholders trained within each company's internal organization that will design the plan and be responsible for executing the future horizontal collaboration, as well as between the necessary facilities to combine operations, must be taken into consideration. The different means of transportation also need to be compatible for effective collaboration as do the logistics conditions for the products, and the technology and computer programs needed must be adequate for sharing information and supporting joint decision making.
4. The need for a **neutral role** in the figure which has been referred to as the facilitator or trustee and does the online and offline work in the collaboration is essential when it comes to overcoming the lack of trust, especially in the beginning, between the potential collaborators. To this end, using the services of a logistics operator which besides offering a global vision of their customers' supply networks and having extensive experience working with different sales mechanisms, could be added as a stakeholder that can objectively assess the benefits that could come out of the collaborative operations for all of the parties especially when the collaboration network grows. Some of the companies that participated in this research work shared the opinion that sometimes companies that use logistics service provider waive control over their transport and thus trust these operators to be the ones to carry out these types of collaboration initiatives.
5. The **means of communicating information** must allow for the exchange of the information under total transparency. The most common means used are telephone and email for the types of collaboration discussed herein. However, although the information shared at the start of a collaborative relationship is limited, the information is exposed with less concern as the relationship grows and a relationship of mutual trust is created. In these cases, other communications methods could be suggested.

6. It is necessary to establish **rules for entering and leaving** the different consortiums. The rules of entry should include the requirements that potential collaborators must fulfill in order to enter an already existing coalition. As far as the rules of departure, and as previously mentioned, they must be explicitly reflected with the inclusion of an exit clause in the agreements signed so as to ensure the sustainability of the collaboration.
7. The same is true for the transfer of relevant information as the need to sign contracts grows as the scope of the horizontal collaboration grows. Although it may seem that signing a contract goes against the essence of horizontal collaboration based on mutual trust and work, it is considered necessary to **have some type of legal framework** that establishes beforehand and in writing aspects as relevant as the responsibilities assumed by each partner including, for example, who shall ensure the loading, how the benefits will be distributed or what the rules are for entering and leaving the consortium. In developing this research work, the difficulty of agreeing to the terms of the very horizontal collaboration was noted meaning these agreements must establish the operational aspects a priori.
8. **A balanced and fair as possible distribution of the benefits** (qualitative and quantitative) is also a fundamental aspect to keep in mind which will make or break the collaboration over time. There are many methods for establishing this distribution of the payoffs using an equitable benefit method, game theory using Shapley Value (Shapley 1953), etc. The latter was proposed in this study as the most adequate and just method for distributing the savings generated; the procedures for calculating it were also offered. The distribution method to be used, which may be the one proposed herein or one the collaborators deem more appropriate in their particular case, should be agreed to and included in the agreements signed prior to the start of operations. Correcting these quantitative distributions by recognizing other qualitative aspects which acknowledge each member's efforts in the collaboration such as a greater effort of one of the partners in modifying its individual baseline operations or one partner's advice on certain logistics aspects is also recommended.

References

Heide JB, John G (1992) Do norms matter in marketing relationships? J Market 56(2):32–44

Lado AA, Dant RR, Tekleab AG (2008) Trust-opportunism paradox, relationalism and performance in interfirm relationships: evidence from the retail industry. Strateg Manag J 29:401–423

Lambert DM, Emmelhainz MA, Gardner TJ (1999) Building successful logistics partnerships. J Bus Logis 20(1):165–181

Shapley LS (1953) A value for n-person games. In: Kuhn HW, Tucker AW (eds) Contributions to the theory of games, vol II. Princeton University Press. Ann Math Stud 28:307–317

Annexes

Annex I: Interview Guide

Introduction

Thank you for taking your time out to meet with us. I would like to take a few minutes to explain the research project to you. The mission of CO_3 is to force a structural breakthrough in the competitiveness and sustainability of European logistics by stimulating horizontal collaboration between European shippers. To achieve this goal, horizontal flow bundling and co-modality scenarios, designed to be repeatable and scalable, will be created in a number of test cases. To facilitate this, a number of company interviews are taking place to assess the attitudes towards collaborative partnerships. The focus is on horizontal collaboration and the results of the interviews will help identify suitable business models for successful collaborations.

The Interview for Horizontal Collaboration

THE COMPANY

1. General information (size, sector, products)
2. Competitive environment
3. Network description: volume, economies of scale, distribution levels
4. Horizontal Collaboration Opportunities within the Supply Chain

COMPANY CULTURE

5. Competitive strategy (differentiation)
6. Sustainability
7. Attitudes towards collaboration

COLLABORATION

8. Background/General collaborative experiences
9. Horizontal Collaboration experiences

M.J. Saenz et al., *Enabling Horizontal Collaboration Through Continuous Relational Learning*, SpringerBriefs in Operations Research,
DOI 10.1007/978-3-319-08093-2

Drivers for Collaboration[1]:

10. Cost reduction
11. Allowing easier response to demand fluctuation
12. Improvement of the service level
13. Improvement of the vehicle fill utilisation
14. Lower carbon emissions
15. Access new markets

Most important barriers to begin collaboration

16. Organisational Culture
17. Lack of trust
18. Difficulty in finding partners
19. Lack of common processes
20. Competitors gaining information
21. Difficulty agreeing terms of HC
22. Difficulty in planning value sharing

ABOUT THE HORIZONTAL COLLABORATION

Exploration

23. What was the main driver to start the collaboration?
24. Did you make a previous analysis of the partner's company?
25. Why them?
26. How were the first contacts managed?
27. How have those changed along time (frequency, upstream/downstream spread within the organization)?
28. Initial evaluation of the following components:

Relational components

- Trust (honesty, sincerity and security)
- Mutuality (fair balance in terms of cost and benefit sharing from the collaboration)
- Solidarity (both companies help each other out when facing difficult situations)
- Relational flexibility (capability to make changes in the company for the benefit of a common goal)
- Organizational compatibility (in terms of corporate culture)

Operational components

- Operative symmetry (level of technical compatibility, demand, logistic network, etc.)
- Dependency of one's Logistics Operator

[1] Selected points have Likert scale options where 1 is low or unlikely and 5 is high or definitely.

29. What kind of collaboration was meant to take place at first?
30. Was there a conviction since the very beginning of the necessity to write a contract for the collaboration?

Assimilation

31. Did the exploration change your attitude towards collaboration?
32. Do you foresee doing technological (e.g. software), methodological or processes changes to enable the collaboration?
33. Do you foresee doing any kind of selection and/or training action of your Human Resources to manage the collaboration?
34. How did your crew assimilate working with the other company's crew?

Exploitation

35. In what way has the exploitation changed your company's opinion about the efficiency of collaborative actions?
36. The technological and managerial changes made by your organization where appropriate and sufficient for the collaboration?
37. Did your distribution network change?
38. Did any of the partners change their logistics operator due to the collaboration?
39. Did the logistics operator have to change their operations due to the collaboration?
40. Have there been any changes on the modal shifts due to the collaboration?
41. What proportion of your deliveries makes part of the common shipments?
42. Did you have to make changes on your packaging design?
43. In what way has the collaboration change your picking operations?
44. Have there been any changes in the frequency of your delivery?
45. Evolution from the initial evaluation of the following components:

    Relational components

    - Trust (honesty, sincerity and security)
    - Mutuality (fair balance in terms of cost and benefit sharing from the collaboration)
    - Solidarity (both companies help each other out when facing difficult situations)
    - Relational flexibility (capability to make changes in the company for the benefit of a common goal)
    - Organizational compatibility (in terms of corporate culture)

Operational components

- Operative symmetry (level of technical compatibility, demand, logistic network, etc.)
- Dependency of one's Logistics Operator

Contracts:

a. Terms
b. Penalties
c. Gain sharing: sharing of the benefits and added value

Contact:

a. At what level and with which frequency are the contacts for strategic supervision set up?
b. Which persons and in what way do they manage the joint distribution operations?
c. Did the contact with your partner open a way of joint learning beyond the pure logistic operations?

46. Roles of the logistics operator, trustee and orchestrator

Benefits of the collaboration $\Longrightarrow$ how do you expect or have you been able to improve your:

a. Efficiency
b. Flexibility
c. Service level
d. Synergy
e. Innovation
f. Sustainability

Annex II: Data Gathering for Assessing Horizontal Collaboration

1. COMPANY DATA

1.1. Name
1.2. Sector/subsector
1.3. Person for contact
1.4. Role of the person for contact
1.5. Telephone number of the person for contact
1.6. E-mail of the person for contact
1.7. Size (annual invoicing and number of employees)
1.8. Network description: Local, National and/or International
1.9. % of sales in local, national and international market
1.10. Countries (name the 5 countries with major exportation volume)
1.11. Function of this size in the Company network
1.12. Picking in Zaragoza? Volume? Number of costumers? Own means or external services?
1.13. Seasonal or continuous production?
1.14. Which companies are you willing to collaborate with?
1.15. Other comments:________________________

2. DELIVERY NETWORK (FOR EACH CUSTOMER)

2.1. Name
2.2. Type
2.3. Origins of delivery route
    2.3.1. Country
    2.3.2. City
    2.3.3. Postal Code
    2.3.4. Own warehouse or external services?
2.4. Destiny of delivery route
    2.4.1. Country
    2.4.2. City
    2.4.3. Postal Code
    2.4.4. Own warehouse or external services?
2.5. Frequency of delivery
    2.5.1. Daily, weekly, monthly,…
    2.5.2. Which days?
2.6. Delivery date variability (indicate the temporal window that the costumer allows)
2.7. Special Requirements of transport
2.8. Transport mode

2.9. Own transport means or External services? (In the second case indicate name and data of the logistics services provider)
2.10. Order composition (full truck load or partially load)
2.11. Delivery format (cases, full pallets,…)
2.12. Quantity per delivery (kgs, Tn, number of cases, number of pallets,…)
2.13. Routes planning preparation (days)
2.14. Information systems
2.15. KPI's used
2.16. % storage costs in total product cost
2.17. % transport costs in total product cost
2.18. Possible returns exploitation?

3. SUPPLY NETWORK (FOR EACH SUPPLIER)

3.1. Name
3.2. Origins of supply route

  3.2.1. Country
  3.2.2. City
  3.2.3. Postal Code

3.3. Destiny of supply route

  3.3.1. Country
  3.3.2. City
  3.3.3. Postal Code

3.4. Frequency of delivery

  3.4.1. Daily, weekly, monthly,…
  3.4.2. Which days?

3.5. Supply date variability (indicate the time window that is allowed to the supplier)
3.6. Special requirements of transport
3.7. Transport mode
3.8. Own transport means or External services? (In the second case indicate name and data of the logistics services provider)
3.9. Order composition (full truck load or partial load)
3.10. Delivery format (cases, full pallets,…)
3.11. Quantity per order (kgs, Tn, number of cases, number of pallets,…)
3.12. Routes planning preparation (days)
3.13. Information systems
3.14. KPI's used
3.15. Possible returns exploitation?

Annex III: Maps Sample

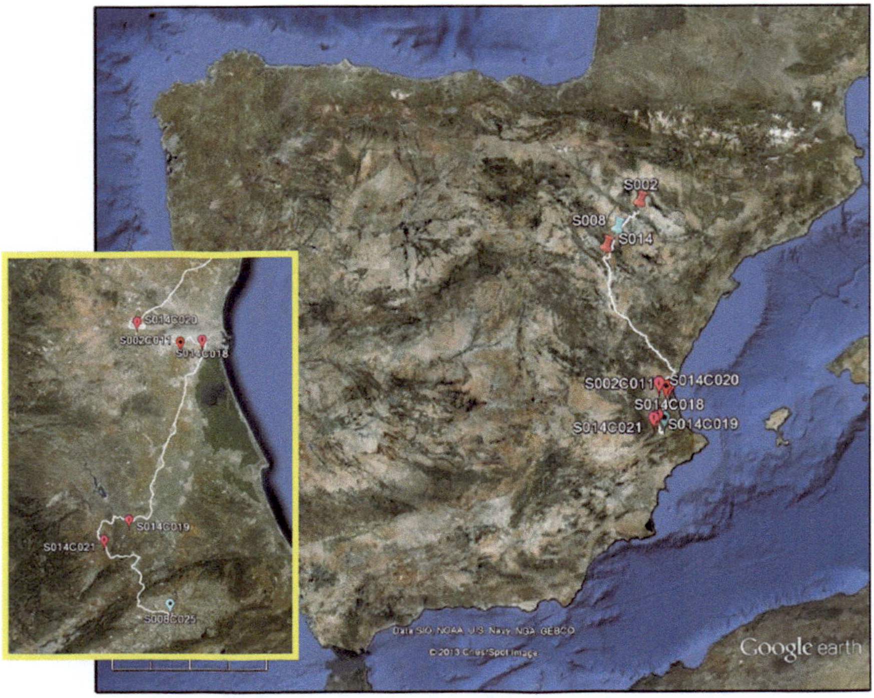
S002
S008
S014
S002C011
S014C020
S014C018
S014C021
S014C019
S014C020
S002C011
S014C018
S014C019
S014C021
S008C025
Data SIO, NOAA, U.S. Navy, NGA, GEBCO
© 2013 Cnes/Spot Image
Google earth

S002
S008
S014
S014C002
S014C014
S008C024
S002C004
Data SIO, NOAA, U.S. Navy, NGA, GEBCO
© 2013 Cnes/Spot Image
Google earth

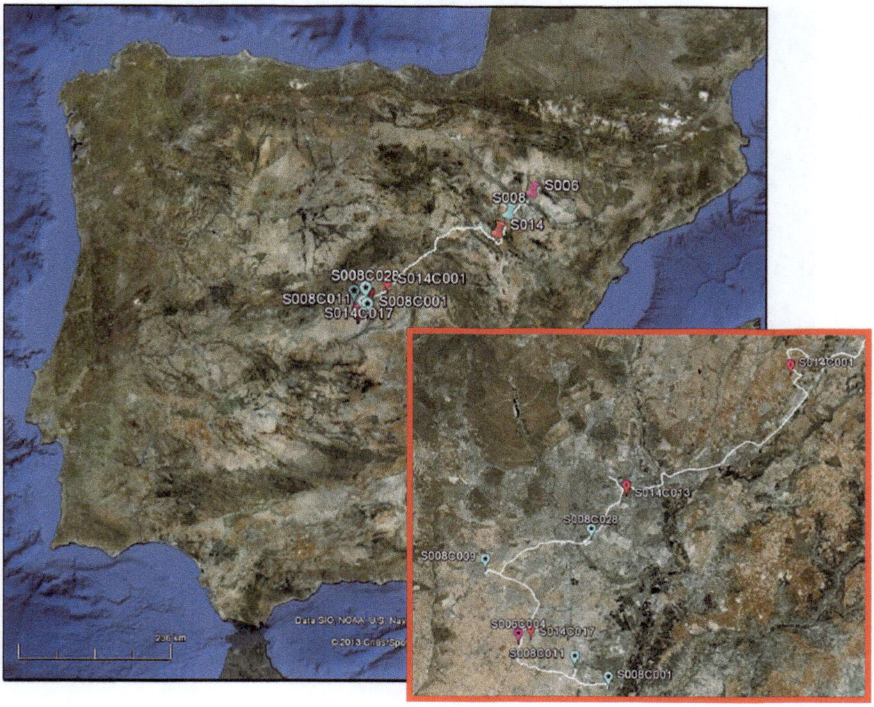
S006
S008
S014
S008C028
S014C001
S008C011
S008C001
S014C017
236 km
S014C013
S008C028
S008C009
S014C017
S008C011
S008C001

S002
S008
S014
S014C010
S002C012
S008C026
S008C018
Data SIO, NOAA, U.S. Navy, NGA, GEBCO
© 2013 Cnes/Spot Image
234 km
Google earth

S006
S014
S008
S014C008
S006C007
S008C020
Data SIO, NOAA, U.S. Navy, NGA, GEBCO
© 2013 Cnes/Spot Image
231 km
Google earth

S002
S006
S014C022
S014
S002C006
S014C016
S006C006
S006C003
S006C005
S002C007
Data SIO, NOAA, U.S. Navy, NGA, GEBCO
© 2013 Cnes/Spot Image
230 km
Google earth

S014C011
S014C003
S002C002
S014C007
S014C015
S014C024
S014C023
S002
S014
Google earth

MIX
Papier aus verantwortungsvollen Quellen
Paper from responsible sources
FSC® C105338

If you have any concerns about our products,
you can contact us on
ProductSafety@springernature.com

In case Publisher is established outside the EU,
the EU authorized representative is:
Springer Nature Customer Service Center GmbH
Europaplatz 3, 69115 Heidelberg, Germany

Printed by Libri Plureos GmbH
in Hamburg, Germany